Theories of Explanation

THEORIES
OF EXPLANATION

Edited by
JOSEPH C. PITT

New York   Oxford
OXFORD UNIVERSITY PRESS
1988

Oxford University Press

Oxford   New York   Toronto
Delhi   Bombay   Calcutta   Madras   Karachi
Petaling Jaya   Singapore   Hong Kong   Tokyo
Nairobi   Dar es Salaam   Cape Town
Melbourne   Auckland

and associated companies in
Beirut   Berlin   Ibadan   Nicosia

Library of Congress Cataloging-in-Publication Data
Theories of explanation.
Includes bibliographies.
1. Science—Philosophy.   2. Explanation (Philosophy)
I. Pitt, Joseph C.
Q175.T485   1988   501   87-7978
ISBN 0-19-504970-5
ISBN 0-19-504971-3 (pbk.)

2 4 6 8 9 7 5 3 1

Printed in the United States of America
on acid-free paper

Contents

Theories of Explanation

1

Introduction

Joseph Pitt

The Success of Science

We study the nature of science for many reasons. Two of the most obvious are: (1) it has been extremely successful, (2) the impact of the fruits of scientific work has dramatically changed the character of contemporary life and promises to contribute to even greater changes in the future. For our purposes, the most important of the reasons listed above is the first. Because of the impact of science on our lives, we need to come to some understanding of why and how it is so successful.

But understanding what makes science successful is a difficult task. For what do we mean when we say science has been successful? We could be referring to the spin-off technologies science is supposed to have generated. Unfortunately, the line from science to technology is not a straight one, and it is not always obvious what the role of science has been with respect to any given technological innovation. Furthermore, there is the largely unexplored problem of the impact of technology on the development of science. We are in the habit of thinking that science leads to technology. But technology also leads to science. This is especially the case in newly emerging areas where technologies developed in conjunction with one science are imported to assist in solving different kinds of problems in another science. As an example, consider the role X-ray crystallography played in the discovery of the structure of DNA. The use of the technology was not developed for purposes of biological research. Rather it has its origins in physics. But it was used to help solve a problem in the foundations of biology, the solution to which has led to the development of yet another technology: genetic engineering. In short, the relationship between the growth of science and the development of technology is not obvious. Until it is clear we cannot use the growth and development of technology as a measure of the success of science.

A second thing we could mean by the success of science is that science has provided us with an increasingly more accurate picture of the pieces and structures of the universe, that is, science has been successful because it has increased our knowledge. And, like the claim that technological innovation is an indicator of the success of science, the view that the success of science can be measured by our increased knowledge also has its problems. For one thing, it is not clear what we mean by "knowledge." If by "knowledge" we mean "increased ability to manipulate the world," then we seem to be stuck in a circle: science is successful because with

3

the development of science we become demonstrably better at manipulating our environment to achieve our goals. But, this amounts to nothing more than the claim that our science is successful because we are successful in manipulating our environment, which tells us nothing about why the science is successful. Furthermore, it assumes the very thing in question: the connection between science and our success in the world. If, on the other hand, by "knowledge" we mean something like "have evidence in support of theoretical claims that the world is populated by such-and-such entities and is governed by so-and-so laws," then we have a different sort of difficulty. The problem here is that we do not have enough control of the evidence-giving relation to know when we are in a position to claim that having evidence entails we are justified in our cognitive claims. There are two sides to this problem: logical and historical.

As regards the logical point of view, in the late twenties and thirties the logical positivists attempted to characterize the nature of scientific knowledge. Initially these philosophers resided primarily in Germany and Austria; with the increasing political problems associated with the rise of nazism and the advent of World War II, they came to reside in Britain and the United States. There their views gained currency, and the position they developed came to dominate the philosophy of science. This influence lasted until the sixties, when, under the pressure of internal problems and faced with an alternative posed by Thomas Kuhn's work, its popularity waned. But to say it waned is not to say it died. Recently it has resurfaced and is making an interesting attempt at a comeback.

The view initially advocated by the logical positivists attempted to articulate precisely the conditions for scientific knowledge. They first proposed a very strong condition: verifiability. One was justified in claiming to know that x if x was verified. Unfortunately, as it turned out, the condition created problems because it was too strong. The initial efforts to make sense of verification gave way to a weaker condition: confirmation. That too had its problems and efforts to resolve them continue to this day. The fundamental issue here is the construction of a strong enough relation between an hypothesis and the evidence for it to allow asserting the hypothesis, that is, to claim it as knowledge. Basically the problem is one of determining what it means to say you have enough evidence. On examination, keeping Hume's problem of induction in mind, it appears any degree of support short of invoking total evidence (all evidence past, present and future) is arbitrary. For unless we can be certain that our meager evidence base, drawn from our spotty investigation of past events, *guarantees* the truth of the prediction that an event x will occur because of what has happened in the past, then evidence alone is not sufficient to claim you know something. To escape this problem a truth condition was invoked. In place of certainty, knowledge was claimed to be justified true belief.

But the truth condition has its own problems. For example, it is not clear when it has been satisfied. It is one thing to say that a cognitive claim counts as knowledge when it is justified (satisfies some methodological constraint or other) and is true, but it is a totally different issue to determine when the claim is in fact true. For some the truth condition is simply another side of the verification condition. What is sought here is some independent means of determining the cognitive status of the claim in question, and the fact of the matter is that form of independent check cannot be had.

Another way to view the development of the logical positivist's account of scientific knowledge is to see it as evolving away from an older and very long tradition, in which knowledge means certainty, toward the newer view, associated with Humean empiricism, that knowledge is justified true belief. But the truth condition is just another way of asking for certainty. To truly differentiate between the two views we need to drop the truth condition. However, if we drop the truth condition and settle for justified belief, the fear is that this will reduce to straightforward belief. We can then pose the problem rather starkly: do we really want to say that knowledge is nothing more than belief? Despite all the efforts to bolster the concept of belief by building in a series of logical safeguards which tie evidence to hypotheses, the feeling remains that there is a difference between knowledge and belief. Even on those accounts which are alternatives to the positivist's, the same problem emerges, for the issue of knowledge versus belief recurs. Once knowledge conceived of as certainty is abandoned for something cognitively less demanding, the issue of whether we ever really have knowledge is with us permanently. So, to the extent that philosophers of science have addressed the problem of knowledge and have settled for something less robust than verification and certainty, the nature of scientific knowledge remains obscure. And if we can't be sure of the nature of scientific knowedge, then we cannot appeal to the increase of knowledge as evidence for the success of science.

The second problem area associated with attempting to measure the success of science by pointing to an increase in our knowledge concerns the history of science. Not surprisingly, there are disagreements over how to read it. These disagreements concern the question of whether or not scientific knowledge is cumulative—that is, whether it grows over time. The logical positivists believed that scientific knowledge *was* cumulative. This followed from their belief that once you were able to produce a truly verified claim, it would be true for all time. On their view science discovered truths about the world and, once discovered, these truths were never discarded. Thus, scientific knowledge represented an accumulation of truths over time. The march of science towards total knowledge was steady and irreversible.

The opposition to the cumulative view of scientific knowledge was crystallized in Thomas Kuhn's *Structure of Scientific Revolutions*. This work contributed greatly to undermining the positivist's assumptions about the growth of knowledge. Kuhn showed that the history of science could be conceptualized as developing by revolutions. During these revolutions the basic assumptions and methods which guided the science of the time (Kuhn's term for this combination of features is 'paradigm') are abandoned in the face of mounting problems in favor of a new paradigm. Swept away with the old paradigm is all the knowledge it generated, since the new paradigm, by definition, uses new methods and assumptions which are incompatible with the old. Thus, on this view, we have something equivalent to pots of knowledge associated with paradigms and when a paradigm is replaced its associated pot of knowledge is thrown out as well. Furthermore, since the various paradigms and pots of knowledge are incompatible, they are not comparable. Hence, it is impossible to determine if one pot has more knowledge in it than another; hence, the impossibility of referring to the success of science by pointing to the increasing amount of knowledge in this pot as opposed to that pot.

In addition to this confusing state of affairs, there is one more difficulty posed by the change-by-revolutions school of thought: since all previous scientific knowledge has been replaced, and since science seems to develop by change of paradigms, doesn't it follow that everything we now know probably will be rejected when the next scientific revolution comes? Even if we reject Kuhn's account and settle for something less dramatic, a version of this problem remains with us. For as soon as we appeal to the history of science for insight into the nature of the growth of scientific knowledge, we cannot help but observe the difference between the theories we now use and those that were prominent a short while ago. It may be too strong to assert that the history of science is the history of the rejection of false theories, but something fairly close is not out of order. The past success of science in discovering truths about the universe dissolves under scrutiny into the continuing failure of science to produce anything solid enough to last unchanged. As new theories are developed, the universal truths of yesteryear are either rejected out of hand or modified or contextualized so as to restrict their scope. Once again it does not look promising for the idea that the success of science can be cashed out in terms of increased knowledge.

So far we have examined some of the approaches philosophers have pursued to account for the success of science. The fact that the efforts of the philosophers have failed does not mean that science has not been successful. It means only that philosophers have been unsuccessful in accounting for the success of science. Furthermore, the lack of philosophical success may be a function of having asked the wrong questions. Given the overwhelming presence of science in contemporary life, the natural temptation is to inquire as to how it has managed to increase our knowledge and foster technological growth. Nevertheless, as we have seen, the answers to these sorts of issues may require more than we can provide by way of answers. In cases like these, when the questions refuse to give way under repeated analysis, it is appropriate to ask if the question has been properly formulated. Queries such as these about the adequacy of the formulation of problems requires an investigation into the assumptions that lead to the initial formulation of the problem.

We have found that some of the questions that trouble philosophers concern the relation between science and technology and the nature of scientific knowledge. These two areas have something in common. It is not by accident that philosophers have focused their concerns about science on these problems. Even though the questions philosophers have asked about these two problems on the surface may often appear quite disparate, there is a common assumption behind them. That assumption as normally invoked may be mistaken, but that is not the issue here. The assumption runs something like this: science is successful because it discovers how the world works and it is that knowledge which makes technological advancement possible. Whether or not science discovers how the world works or how some version of the world works is not the issue. Nor is the fact that it is extremely difficult to see our way from theoretical descriptions of the microstructure of the universe to home computers really a problem here. What is crucial is the insight that the kind of knowledge science produces, whatever form that knowledge takes, permits the development of explanations, and it is those explanations which are the real payoff. Thus, demands for an account of scientific knowledge, however interesting in their own right, may not contribute to our understanding of the success of science. What we need to understand is the nature of scientific explanation.

The Role of Explanation

If the above ruminations are correct, we can now approach the topic at hand. Philosophers worry about the nature of scientific explanation for two reasons: (1) explanations are supposed to tell us how things work and knowing how things work gives us the power to manipulate our environment to achieve our own ends, and (2) science is supposed to be our best means of generating explanations which satisfy the criterion of providing the means to accomplish our goals. It really doesn't matter what "knowledge" means. What matters is what counts as an explanation. By putting the emphasis on explanation we make it a criterion of adequacy for scientific progress. All the research in the world counts for nothing if it fails to generate explanations of the domain under investigation.

To some this may sound like an overly pragmatic characterization of the nature of the scientific enterprise. What, for example, ever happened to the ideal of knowledge for its own sake, or to the disinterested observer, or to the scientist concerned solely with the problem and its solution? From our perspective these are primarily either interesting fictions or irrelevant to the issue at hand, which is to account for the philosophical importance of scientific explanation. There may be disinterested observers, there may even be individuals who have managed to convince themselves that there is something called knowledge for its own sake. But that is not important. Whatever knowledge may be, its hallmark is the ability to do something with it. In the case of scientific knowledge this means offering an explanation for some phenomenon or other. If it can't successfully be used in some such fashion then it doesn't qualify as knowledge.

In defense of my overly pragmatic account of the importance of explanation, let us in passing briefly reflect on the evolution of the importance of the concept of explanation in the philosophy of science in general. It has gone from being one concept among others to the foundation of a theory of justification. Behind all the philosophers' concerns about science lies the notion of justification. The justification of scientific claims is the key problem because without justification our conviction that scientific knowledge is our best example of knowledge (because of the success of science?) fails. With it also fails the ideal of science as our foremost example of rationality. Justification is, therefore, the key to understanding the philosopher's concerns about knowledge in general and science in particular.

The use of explanation as a criterion of adequacy for scientific claims is behind a current theory of justification which regards both justification and explanation differently than the positivists did. The positivists formulated an account of explanation on the basis of a theory of justification: verificationism. Their method was to analyze key concepts of science such as evidence, theory, explanation, law and hypothesis, in the context of the role of a formal language, the axiomatization of theories and the verificationist theory of meaning. As parts of this program failed or became modified other parts were affected, sometimes positively, sometimes not. But where explanation was concerned an interesting turnabout took place. With the failure of the verification/confirmation theory of justification (and I use "failure" here in the weak sense of not meeting the restrictions the positivists themselves laid down) a new theory of justification was proposed based on the concept of explanation. On this new view a cognitive claim was justified if it produced an inference to the best explanation or was deduced from an inference to the best explanation. Thus, the justification for cognitive claims was not to be found in their status as knowledge, or strictly in terms of the evidence supporting them, but rather in terms of their role in explanations. This is not to say that explanationism, as Keith Lehrer called the new theory, is to everyone's liking. Nor is it to suggest that explanationism is "the ruling theory," which it is not. Finally, it is also not the case that it is a relatively new idea, since its use can be seen as far back as Galileo. What is important for our purposes is to see the evolution of our perception of the importance of explanation from one concept *inter alia* whose logical structure needed explication to the use of the notion as the fundamental idea in a theory of justification. The evolution of the idea is, I believe, a function of our willingness to reexamine our fundamental assumptions about justification.

This leaves us with the question of what constitutes an explanation. I do not intend to try to answer that question. That is what the papers in this volume are concerned with. They have been selected to provide the reader with some sense of the emerging scope of the dialogue, beginning with the appearance of the first major work on the topic, Hempel and Oppenheim's "Studies in the Logic of Explanation." This paper is the touchstone for the development of the topic in the contemporary literature. As can be seen from the ensuing discussion, the problem of explanation is far from settled. What is clear is the centrality of its role in our emerging understanding of science. As such it demands our attention.

References

Hempel, C., and Oppenheim, P., "Studies in the Logic of Explanation." *Philosophy of Science*, vol. 15, 1948.

Kuhn, Thomas, *The Structure of Scientific Revolutions*. 2nd rev. ed. Chicago: University of Chicago Press, 1970.

Lehrer, Keith, "Justification, Explanation, and Induction." In *Induction, Acceptance and Rational Belief*. Ed. M. Swain. Dordrecht: Reidel, 1970.

Neurath, O., Carnap, R., and Morris, C., eds., *Foundation of the Unity of Science*, vols. 1 and 2. Chicago: University of Chicago Press, 1970.

Pitt, Joseph, "Galileo, Rationality, and Explanation." *Philosophy of Science*, vol. 55, 1988.

Sellars, Wilfrid, *Science, Perception and Reality*. London: Routledge and Kegan Paul, 1969.

2

Studies in the Logic of Explanation

Carl G. Hempel and Paul Oppenheim

1. Introduction

To explain the phenomena in the world of our experience, to answer the question "why?" rather than only the question "what?" is one of the foremost objectives of empirical science. While there is rather general agreement on this point there exists considerable difference of opinion as to the function and the essential characteristics of scientific explanation. The present essay is an attempt to shed some light on these issues by means of an elementary survey of the basic pattern of scientific explanation and a subsequent more rigorous analysis of the concept of law and the logical structure of explanatory arguments.[1]

The elementary survey is presented in Part I; Part II contains an analysis of the concept of emergence; Part III seeks to exhibit and to clarify in a more rigorous manner some of the peculiar and perplexing logical problems to which the familiar elementary analysis of explanation gives rise. Part IV, finally, deals with the idea of explanatory power of a theory; an explicit definition and a formal theory of this concept are developed for the case of a scientific language of simple logical structure.

Part I. Elementary Survey of Scientific Explanation

2. *Some Illustrations*

A mercury thermometer is rapidly immersed in hot water; there occurs a temporary drop of the mercury column, which is then followed by a swift rise. How is this phenomenon to be explained? The increase in temperature affects at first only the glass tube of the thermometer; it expands and thus provides a larger space for the mercury inside, whose surface therefore drops. As soon as by heat conduction the rise in temperature reaches the mercury, however, the latter expands, and as its coefficient of expansion is considerably larger than that of glass, a rise of the mercury level results. This account consists of statements of two kinds. Those of the first kind indicate certain conditions which are realized prior to, or at the same time as,

C. G. Hempel, P. Oppenheim, Studies in the Logic of Explanation, *Philosophy of Science*, vol. 15, pgs. 567–579, © by Williams & Wilkins, 1948.

the phenomenon to be explained; we shall refer to them briefly as antecedent conditions. In our illustration, the antecedent conditions include, among others, the fact that the thermometer consists of a glass tube which is partly filled with mercury, and that it is immersed into hot water. The statements of the second kind express certain general laws; in our case, these include the laws of the thermic expansion of mercury and of glass, and a statement about the small thermic conductivity of glass. The two sets of statements, if adequately and completely formulated, explain the phenomenon under consideration: they entail the consequence that the mercury will first drop, then rise. Thus, the event under discussion is explained by subsuming it under general laws, that is, by showing that it occurred in accordance with those laws, in virtue of the realization of certain specified antecedent conditions.

Consider another illustration. To an observer in a rowboat, that part of an oar which is under water appears to be bent upwards. The phenomenon is explained by means of general laws—mainly the law of refraction and the law that water is an optically denser medium than air—and by reference to certain antecedent conditions—especially the facts that part of the oar is in the water, part in the air, and that the oar is practically a straight piece of wood. Thus, here again, the question "*Why* does the phenomenon occur?" is construed as meaning "according to what general laws, and by virtue of what antecedent conditions does the phenomenon occur?"

So far, we have considered only the explanation of particular events occurring at a certain time and place. But the question "why?" may be raised also in regard to general laws. Thus, in our last illustration, the question might be asked: Why does the propagation of light conform to the law of refraction? Classical physics answers in terms of the undulatory theory of light, that is, by stating that the propagation of light is a wave phenomenon of a certain general type, and that all wave phenomena of that type satisfy the law of refraction. Thus, the explanation of a general regularity consists in subsuming it under another, more comprehensive regularity, under a more general law. Similarly, the validity of Galileo's law for the free fall of bodies near the earth's surface can be explained by deducing it from a more comprehensive set of laws, namely Newton's laws of motion and his law of gravitation, together with some statements about particular facts, namely, about the mass and the radius of the earth.

3. The Basic Pattern of Scientific Explanation

From the preceding sample cases let us now abstract some general characteristics of scientific explanation. We divide an explanation into two major constituents, the *explanandum* and the *explanans*.[2] By the explanandum, we understand the sentence describing the phenomenon to be explained (not that phenomenon itself); by the explanans, the class of those sentences which are adduced to account for the phenomenon. As was noted before, the explanans falls into two subclasses; one of these contains certain sentences $C_1, C_2, \ldots, C_k$ which state specific antecedent conditions; the other is a set of sentences $L_1, L_2, \ldots, L_r$ which represent general laws.

If a proposed explanation is to be sound, its constituents have to satisfy certain conditions of adequacy, which may be divided into logical and empirical conditions. For the following discussion, it will be sufficient to formulate these requirements in a slightly vague manner; in Part III, a more precise restatement of these criteria will be presented.

I. *Logical Conditions of Adequacy*

(R1) The explanandum must be a logical consequence of the explanans; in other words, the explanandum must be logically deducible from the information contained in the explanans; for otherwise, the explanans would not constitute adequate grounds for the explanandum.

(R2) The explanans must contain general laws, and these must actually be required for the derivation of the explanandum. We shall not make it a necessary condition for a sound explanation, however, that the explanans must contain at least one statement which is not a law; for, to mention just one reason, we would surely want to consider as an explanation the derivation of the general regularities governing the motion of double stars from the laws of celestial mechanics, even though all the statements in the explanans are general laws.

(R3) The explanans must have empirical content; that is, it must be capable, at least in principle, of test by experiment or observation. This condition is implicit in (R1); for since the explanandum is assumed to describe some empirical phenomenon, it follows from (R1) that the explanans entails at least one consequence of empirical character, and this fact confers upon it testability and empirical content. But the point deserves special mention because, as will be seen in §4, certain arguments which have been offered as explanations in the natural and in the social sciences violate this requirement.

II. *Empirical Condition of Adequacy*

(R4) The sentences constituting the explanans must be true. That in a sound explanation, the statements constituting the explanans have to satisfy some condition of factual correctness is obvious. But it might seem more appropriate to stipulate that the explanans has to be highly confirmed by all the relevant evidence available rather than that it should be true. This stipulation, however, leads to awkward consequences. Suppose that a certain phenomenon was explained at an earlier stage of science, by means of an explanans which was well supported by the evidence then at hand, but which has been highly disconfirmed by more recent empirical findings. In such a case, we would have to say that originally the explanatory account was a correct explanation, but that it ceased to be one later, when unfavorable evidence was discovered. This does not appear to accord with sound common usage, which directs us to say that on the basis of the limited initial evidence, the truth of the explanans, and thus the soundness of the explanation, had been quite probable, but that the ampler evidence now available makes it highly probable that the explanans is not true, and hence that the account in question is not—and never has been—a correct explanation.[3] (A similar point will be made and illustrated, with respect to the requirement of truth for laws, in the beginning of §6.)

Some of the characteristics of an explanation which have been indicated so far may be summarized in the following schema:

$$\text{Logical deduction} \left[\underbrace{\begin{cases} C_1, C_2, \ldots, C_k & \text{Statements of antecedent conditions} \\ L_1, L_2, \ldots, L_r & \text{General Laws} \end{cases}}_{} \right\} \text{Explanans}$$

$$\left. \begin{array}{c} \quad\quad\quad E \qquad\qquad \text{Description of the empirical phenomenon to be explained} \end{array} \right\} \text{Explanandum}$$

Let us note here that the same formal analysis, including the four necessary conditions, applies to scientific prediction as well as to explanation. The difference between the two is of a pragmatic character. If E is given, that is, if we know that the phenomenon described by E has occurred, and a suitable set of statements C_1, $C_2 \ldots, C_k, L_1, L_2, \ldots, L_r$ is provided afterwards, we speak of an explanation of the phenomenon in question. If the latter statements are given and E is derived prior to the occurrence of the phenomenon it describes, we speak of a prediction. It may be said, therefore, that an explanation of a particular event is not fully adequate unless its explanans, if taken account of in time, could have served as a basis for predicting the event in question. Consequently, whatever will be said in this article concerning the logical characteristics of explanation or prediction will be applicable to either, even if only one of them should be mentioned.[4]

Many explanations which are customarily offered, especially in prescientific discourse, lack this potential predictive force, however. Thus, we may be told that a car turned over on the road "because" one of its tires blew out while the car was traveling at high speed. Clearly, on the basis of just this information, the accident could not have been predicted, for the explanans provides no explicit general laws by means of which the prediction might be effected, nor does it state adequately the antecedent conditions which would be needed for the prediction. The same point may be illustrated by reference to W. S. Jevons's view that every explanation consists in pointing out a resemblance between facts, and that in some cases this process may require no reference to laws at all and "may involve nothing more than a single identity, as when we explain the appearance of shooting stars by showing that they are identical with portions of a comet."[5] But clearly, this identity does not provide an explanation of the phenomenon of shooting stars unless we presuppose the laws governing the development of heat and light as the effect of friction. The observation of similarities has explanatory value only if it involves at least tacit reference to general laws.

In some cases, incomplete explanatory arguments of the kind here illustrated suppress parts of the explanans simply as "obvious"; in other cases, they seem to involve the assumption that while the missing parts are not obvious, the incomplete explanans could at least, with appropriate effort, be so supplemented as to make a strict derivation of the explanandum possible. This assumption may be justifiable in some cases, as when we say that a lump of sugar disappeared "because" it was put into hot tea, but it surely is not satisfied in many other cases. Thus, when certain peculiarities in the work of an artist are explained as outgrowths of a specific type of neurosis, this observation may contain significant clues, but in general it does not afford a sufficient basis for a potential prediction of those peculiarities. In cases of this kind, an incomplete explanation may at best be considered as indicating some positive correlation between the antecedent conditions adduced and the type of phenomenon to be explained, and as pointing out a direction in which further research might be carried on in order to complete the explanatory account.

The type of explanation which has been considered here so far is often referred to as causal explanation.[6] If E describes a particular event, then the antecedent circumstances described in the sentences $C_1, C_2, \ldots, C_k$ may be said jointly to "cause" that event, in the sense that there are certain empirical regularities, expressed by the laws $L_1, L_2, \ldots, L_r$, which imply that whenever conditions of the kind indicated by $C_1, C_2, \ldots, C_k$ occur, an event of the kind described in E will take place. Statements such as $L_1, L_2, \ldots, L_r$, which assert general and unexceptional connections between specified characteristics of events, are customarily called causal, or deterministic, laws. They must be distinguished from the so-called statistical laws which assert that in the long run, an explicitly stated percentage of all cases satisfying a given set of conditions are accompanied by an event of a certain specified kind. Certain cases of scientific explanation involve "subsumption" of the explanandum under a set of laws of which at least some are statistical in character. Analysis of the peculiar logical structure of that type of subsumption involves difficult special problems. The present essay will be restricted to an examination of the deductive type of explanation, which has retained its significance in large segments of contemporary science, and even in some areas where a more adequate account calls for reference to statistical laws.[7]

4. Explanation in the Nonphysical Sciences. Motivational and Teleological Approaches

Our characterization of scientific explanation is so far based on a study of cases taken from the physical sciences. But the general principles thus obtained apply also outside this area.[8] Thus, various types of behavior in laboratory animals and in human subjects are explained in psychology by subsumption under laws or even general theories of learning or conditioning; and while frequently the regularities invoked cannot be stated with the same generality and precision as in physics or chemistry, it is clear at least that the general character of those explanations conforms to our earlier characterization.

Let us now consider an illustration involving sociological and economic factors. In the fall of 1946, there occurred at the cotton exchanges of the United States a price drop which was so severe that the exchanges in New York, New Orleans, and Chicago had to suspend their activities temporarily. In an attempt to explain this occurrence, newspapers traced it back to a large-scale speculator in New Orleans who had feared his holdings were too large and had therefore begun to liquidate his stocks; smaller speculators had then followed his example in a panic and had thus touched off the critical decline. Without attempting to assess the merits of the argument, let us note that the explanation here suggested again involves statements about antecedent conditions and the assumption of general regularities. The former include the facts that the first speculator had large stocks of cotton, that there were smaller speculators with considerable holdings, that there existed the institution of the cotton exchanges with their specific mode of operation, etc. The general regularities referred to are—as often in semipopular explanations—not explicitly mentioned; but there is obviously implied some form of the law of supply and demand to account for the drop in cotton prices in terms of the greatly increased supply under conditions of practically unchanged demand; besides, reliance is necessary on certain regularities in the behavior of individuals who are trying to preserve or improve their economic position. Such laws cannot be formulated at present with satisfactory precision and generality, and therefore, the suggested explanation is surely incomplete, but its intention is unmistakably to account for the phenomenon by integrating it into a general pattern of economic and sociopsychological regularities.

We turn to an explanatory argument taken from the field of linguistics.[9] In Northern France, there are in use a large variety of words synonymous with the English "bee," whereas in Southern France, essentially only one such word is in existence. For this discrepancy, the explanation has been suggested that in the Latin epoch, the South of France used the word "*apicula*," the North the word "*apis*." The latter, because of a process of phonologic decay in Northern France, became the monosyllabic word "é"; and monosyllables tend to be eliminated, especially if they contain few consonantic elements, for they are apt to give rise to misunderstandings. Thus, to avoid confusion, other words were selected. But "*apicula*," which was reduced to "*abelho*," remained clear enough and was retained, and finally it even entered into the standard language, in the form "*abeille*." While the explanation here described is incomplete in the sense characterized in the previous section, it clearly exhibits reference to specific antecedent conditions as well as to general laws.[10]

While illustrations of this kind tend to support the view that explanation in biology, psychology, and the social sciences has the same structure as in the physical sciences, the opinion is rather widely held that in many instances, the causal type of explanation is essentially inadequate in fields other than physics and chemistry, and especially in the study of purposive behavior. Let us examine briefly some of the reasons which have been adduced in support of this view.

One of the most familar among them is the idea that events involving the activities of humans singly or in groups have a peculiar uniqueness and irrepeatability which makes them inaccessible to causal explanation because the latter, with its reliance upon uniformities, presupposes repeatability of the phenomena under consideration. This argument which, incidentally, has also been used in support of the contention that the experimental method is inapplicable in psychology and the social sciences, involves a misunderstanding of the logical character of causal explanation. Every individual event, in the physical sciences no less than in psychology or the social sciences, is unique in the sense that it, with all its peculiar characteristics, does not repeat itself. Nevertheless, individual events may conform to, and thus be explainable by means of, general laws of the causal type. For all that a causal law asserts is that any event of a specified kind, that is, any event having certain specified characteristics, is accompanied by another event which in turn has certain specified characteristics; for example, that in any event involving friction, heat is developed. And all that is needed for the testability and applicability of such laws is the recurrence of events with the antecedent characteristics, that is, the repetition of those characteristics, but not of their individual instances. Thus, the argument is inconclusive. It gives occasion, however, to emphasize an important point concerning our earlier analysis: When we spoke of the explanation of a single event, the term "event" referred to the occurrence of some more or less complex characteristic in a specific spatio-temporal location or in a certain individual object, and not to *all* the characteristics of that object, or to all that goes on in that space-time region.

A second argument that should be mentioned here[11] contends that the establishment of scientific generalizations—and thus of explanatory principles—for human behavior is impossible because the reactions of an individual in a given situation depend not only upon that situation, but also upon the previous history of the individual. But surely, there is no *a priori* reason why generalizations should not be attainable which take into account this dependence of behavior on the past history of the agent. That indeed the given argument "proves" too much, and is therefore a *non sequitur*, is made evident by the existence of certain physical phenomena, such as magnetic hysteresis and elastic fatigue, in which the magnitude of a specific physical effect depends upon the past history of the system involved, and for which nevertheless certain general regularities have been established.

A third argument insists that the explanation of any phenomenon involving purposive behavior calls for reference to motivations and thus for teleological rather than causal analysis. For example, a fuller statement of the suggested explanation for the break in the cotton prices would have to indicate the large-scale speculator's motivations as one of the factors determining the event in question. Thus, we have to refer to goals sought; and this, so the argument runs, introduces a type of explanation alien to the physical sciences. Unquestionably, many of the—frequently incomplete—explanations which are offered for human actions involve reference to goals and motives; but does this make them essentially different from the causal explanations of physics and chemistry? One difference which suggests itself lies in the circumstance that in motivated behavior, the future appears to affect the present in a manner which is not found in the causal explanations of the physical sciences. But clearly, when the action of a person is motivated, say, by the desire to reach a certain objective, then it is not the as yet unrealized future event of attaining that goal which can be said to determine his present behavior, for indeed the goal may never be actually reached; rather—to put it in crude terms—it is (a) his desire, present before the action, to attain that particular objective, and (b) his belief, likewise present before the action, that such and such a course of action is most likely to have the desired effect. The determining motives and beliefs, therefore, have to be classified among the antecedent conditions of a motivational explanation, and there is no formal difference on this account between motivational and causal explanation.

Neither does the fact that motives are not accessible to direct observation by an outside observer constitute an essential difference between the two kinds of explanation; for the determining factors adduced in physical explanations also are very frequently inaccessible to direct observation. This is the case, for instance, when opposite electric charges are adduced in explanation of the mutual attraction of two metal spheres. The presence of those charges, while including direct observation, can be ascertained by various kinds of indirect test, and that is sufficient to guarantee the empirical character of the explanatory statement. Similarly, the presence of certain motivations may be ascertainable only by indirect methods, which may include reference to linguistic utterances of the subject in question, slips of pen or tongue, etc.; but as long as these methods are "operationally determined" with reasonable clarity and precision, there is no essential difference in this respect between motivational explanation and causal explanation in physics.

A potential danger of explanation by motives lies in the fact that the method lends itself to the facile construction of *ex post facto* accounts without predictive force. An action is often explained by attributing it to motives conjectured only after the action has taken place. While this procedure is not in itself objectionable, its soundness, requires that (1) the motivational assumptions in question be capable of test, and (2) that suitable general laws be available to lend explanatory power to the assumed motives. Disregard of these requirements frequently deprives alleged motivational explanations of their cognitive significance.

The explanation of an action in terms of the agent's motives is sometimes considered as a special kind of teleological explanation. As was pointed out above, motivational explanation, if adequately formulated, conforms to the conditions for causal explanation, so that the term "teleological" is a misnomer if it is meant to imply either a non-causal character of the explanation or a peculiar determination of the present by the future. If this is borne in mind, however, the term "teleological" may be viewed, in this context, as referring to causal explanations in which some of the antecedent conditions are motives of the agent whose actions are to be explained.[12]

Teleological explanations of this kind have to be distinguished from a much more sweeping type, which has been claimed by certain schools of thought to be indispensable especially in biology. It consists in explaining characteristics of an organism by reference to certain ends or purposes which the characteristics are said to serve. In contradistinction to the cases examined before, the ends are not assumed here to be consciously or subconsciously pursued by the organism in question. Thus, for the phenomenon of mimicry, the explanation is sometimes offered that it serves the purpose of protecting the animals endowed with it from detection by its pursuers and thus tends to preserve the species. Before teleological hypotheses of this kind can be appraised as to their potential explanatory power, their meaning has to be clarified. If they are intended somehow to express the idea that the purposes they refer to are inherent in the design of the universe, then clearly they are not capable of empirical test and thus violate the requirement (R3) stated in §3. In certain cases, however, assertions about the purposes of biological characteristics may be translatable into statements in non-teleological terminology which assert that those characteristics function in a specific manner which is essential to keeping the organism alive or to preserving the species.[13] An attempt to state precisely what is meant by this latter assertion—or by the similar one that without those characteristics, and other things being equal, the organism or the species would not survive—encounters considerable difficulties. But these need not be discussed here. For even if we assume that biological statements in teleological form can be adequately translated into descriptive statements about the life-preserving function of certain biological characteristics, it is clear that (1) the use of the concept of purpose is not essential in these contexts, since the term "purpose" can be completely eliminated from the statements in question, and (2) teleological assumptions, while now endowed with empirical content, cannot serve as explanatory principles in the customary contexts. Thus, for example, the fact that a given species of butterfly displays a particular kind of coloring cannot be inferred from—and therefore cannot be explained by means of—the statement that this type of coloring has the effect of protecting the butterflies from detection by pursuing birds, nor can the presence of red corpuscles in the human blood be inferred from the statement that those corpuscles have a specific function in assimilating oxygen and that this function is essential for the maintenance of life.

One of the reasons for the perseverance of teleological considerations in biology probably lies in the fruitfulness of the teleological approach as a heuristic device: Biological research which was psychologically motivated by a teleological orientation, by an interest in purposes in nature, has frequently led to important results which can be stated in nonteleological terminology and which increase our knowledge of the causal connections between biological phenomena.

Another aspect that lends appeal to teleological considerations is their anthropomorphic character. A teleological explanation tends to make us feel that we really "understand" the phenomenon in question, because it is accounted for in terms of purposes, with which we are familiar from our own experience of purposive behavior. But it is important to distinguish here understanding in the psychological sense of a feeling of empathic familiarity from understanding in the theoretical, or cognitive, sense of exhibiting the phenomenon to be explained as a special case of some general regularity. The frequent insistence that explanation means the reduction of something unfamiliar to ideas or experiences already familiar to us is indeed misleading. For while some scientific explanations do have this psychological effect, it is by no means universal: The free fall of a physical body may well be said to be a more familiar phenomenon than the law of gravitation, by means of which it can be explained; and surely the basic ideas of the theory of relativity will appear to many to be far less familiar than the phenomena for which the theory accounts.

"Familiarity" of the explanans is not only not necessary for a sound explanation, as has just been noted; it is not sufficient either. This is shown by the many cases in which a proposed explanans sounds suggestively familiar, but upon closer inspection proves to be a mere metaphor, or to lack testability, or to include no general laws and therefore to lack explanatory power. A case in point is the neovitalistic attempt to explain biological phenomena by reference to an entelechy or vital force. The crucial point here is not—as is sometimes said—that entelechies cannot be seen or otherwise directly observed; for that is true also of gravitational fields, and yet, reference to such fields is essential in the explanation of various physical phenomena. The decisive difference between the two cases is that the physical explanation provides (1) methods of testing, albeit indirectly, assertions about gravitational fields, and (2) general laws concerning the strength of gravitational fields, and the behavior of objects moving in them. Explanations by entelechies satisfy the analogue of neither of these two conditions. Failure to satisfy the first condition represents a violation of (R3); it renders all statements about entelechies inaccessible to empirical test and thus devoid of empirical meaning. Failure to comply with the second condition involves a violation of (R2). It deprives the concept of entelechy of all explanatory import; for explanatory power never resides in a concept, but always in the general laws in which it functions. Therefore, notwithstanding the feeling of familiarity it may evoke, the neovitalistic account cannot provide theoretical understanding.

The preceding observations about familiarity and understanding can be applied, in a similar manner, to the view held by some scholars that the explanation, or the understanding, of human actions requires an empathic understanding of the personalities of the agents.[14] This understanding of another person in terms of one's own psychological functioning may prove a useful heuristic device in the search for general psychological principles which might provide a theoretical explanation; but the existence of empathy on the part of the scientist is neither a necessary nor a sufficient condition for the explanation, or the scientific understanding, of any human action. It is not necessary, for the behavior of psychotics or of people belonging to a culture very different from that of the scientist may sometimes be explainable and predictable in terms of general principles even though the scientist who establishes or applies those principles may not be able to understand his subjects empathically. And empathy is not sufficient to guarantee a sound explanation, for a strong feeling of empathy may exist even in cases where we completely misjudge a given personality. Moreover, as Zilsel has pointed out, empathy leads with ease to incompatible results; thus, when the population of a town has long been subjected to heavy bombing attacks, we can understand, in the empathic sense, that its morale should have broken down completely, but we can understand with the same case also that it should have developed a defiant spirit of resistance. Arguments of this kind often appear quite convincing; but they are of an *ex post facto* character and lack cognitive significance unless they are supplemented by testable explanatory principles in the form of laws or theories.

Familiarity of the explanans, therefore, no matter whether it is achieved through the use of teleological terminology, through neovitalistic metaphors, or through other means, is no indication of the cognitive import and the predictive force of a proposed explanation. Besides, the extent to which an idea will be considered as familiar varies from person to person and from time to time, and a psychological factor of this kind certainly cannot serve as a standard in assessing the worth of a proposed explanation. The decisive requirement for every sound explanation remains that it subsume the explanandum under general laws.

Part II. On the Idea of Emergence

5. *Levels of Explanation. Analysis of Emergence*

As has been shown above, a phenomenon may be explainable by sets of laws of different degrees of generality. The changing positions of a planet, for example, may be explained by subsumption under Kepler's laws, or by derivation from the far more comprehensive general law of gravitation in combination with the laws of motion, or finally by deduction from the general theory of relativity, which explains—and slightly modifies—the preceding set of laws. Similarly, the expansion of a gas with rising temperature at constant pressure may be explained by means of the Gas Law or by the more comprehensive kinetic theory of heat. The latter explains the Gas Law, and thus indirectly the phenomenon just mentioned, by means of (1) certain assumptions concerning the micro-behavior of gases (more specifically, the distributions of locations and speeds of the gas molecules) and (2) certain macro-micro principles, which connect such macro-characteristics of a gas as its temperature, pressure, and volume with the micro-characteristics just mentioned.

In the sense of these illustrations, a distinction is frequently made between various *levels of explanation*.[15] Subsumption of a phenomenon under general laws directly connecting observable characteristics represents the first level; higher levels require the use of more or less abstract theoretical constructs which function in the context of some comprehensive theory. As the preceding illustrations show, the concept of higher-level explanation covers procedures of rather different character; one of the most important among them consists in explaining a class of phenomena by means of a theory concerning their micro-structure. The kinetic theory of heat, the atomic theory of matter, the electromagnetic as well as the quantum theory of light, and the gene theory of heredity are examples of this method. It is often felt that only the discovery of a micro-theory affords real scientific understanding of any type of phenomenon, because only it gives us insight into the inner mechanism of the phenomenon, so to speak. Consequently, classes of events for which no micro-theory was available have frequently been viewed as not actually understood; and concern with the theoretical status of phenomena which are unexplained in this sense may be considered as one of the roots of the doctrine of emergence.

Generally speaking, the concept of *emergence* has been used to characterize certain phenomena as "novel," and this not merely in the psychological sense of being unexpected,[16] but in the theoretical sense of being unexplainable, or unpredictable, on the basis of information concerning the spatial parts or other constituents of the systems in which the phenomena occur, and which in this context are often referred to as "wholes." Thus, for example, such characteristics of water as its transparence and liquidity at room temperature and atmospheric pressure, or its ability to quench thirst have been considered as emergent on the ground that they could not possibly have been predicted from a knowledge of the properties of its chemical constituents, hydrogen and oxygen. The weight of the compound, on the contrary, has been said not to be emergent because it is a mere "resultant" of its components and could have been predicted by simple addition even before the compound had been formed. The conceptions of explanation and prediction which underly this idea of emergence call for various critical observations, and for corresponding changes in the concept of emergence.

(1) First, the question whether a given characteristic of a "whole," w, is emergent or not cannot be significantly raised until it has been stated what is to be understood by the parts or constituents of w. The volume of a brick wall, for example, may be inferable by addition from the volumes of its parts if the latter are understood to be the component bricks, but it is not so inferable from the volumes of the molecular components of the wall. Before we can significantly ask whether a characteristic W of an object w is emergent, we shall therefore have to state the intended meaning of the term "part of." This can be done by defining a specific relation Pt and stipulating that those and only those objects which stand in Pt to w count as parts of constituents of w. "Pt" might be defined as meaning "constituent brick of" (with respect to buildings), or "molecule contained in" (for any physical object), or "chemical element contained in" (with respect to chemical compounds, or with respect to any material object), or "cell of" (with respect to organisms), etc. The term "whole" will be used here without any of its various connotations, merely as referring to any object w to which others stand in the specified relation Pt. In order to emphasize the dependence of the concept of part upon the definition of the relation Pt in each case, we shall sometimes speak of Pt-parts, to refer to parts as determined by the particular relation Pt under consideration.

(2) We turn to a second point of criticism. If a characteristic of a whole is counted as emergent simply if its occurrence cannot be inferred from a knowledge of all the properties of its parts, then, as Grelling has pointed out, no whole can

have any emergent characteristics. Thus, to illustrate by reference to our earlier example, the properties of hydrogen include that of forming, if suitably combined with oxygen, a compound which is liquid, transparent, etc. Hence, the liquidity, transparence, etc., of water *can* be inferred from certain properties of its chemical constituents. If the concept of emergence is not to be vacuous, therefore, it will be necessary to specify in every case a class G of attributes and to call a characteristic W of an object w emergent relatively to G and Pt if the occurrence of W in w cannot be inferred from a complete characterization of all the Pt-parts with respect to the attributes contained in G, that is, from a statement which indicates, for every attribute in G, to which of the parts of w it applies. Evidently, the occurrence of a characteristic may be emergent with respect to one class of attributes and not emergent with respect to another. The classes of attributes which the emergentists have in mind, and which are usually not explicitly indicated, will have to be construed as nontrivial, that is, as not logically entailing the property of each constituent of forming, together with the other constituents, a whole with the characteristics under investigation. Some fairly simple cases of emergence in the sense so far specified arise when the class G is restricted to certain simple properties of the parts, to the exclusion of spatial or other relations among them. Thus, the electromotive force of a system of several electric batteries cannot be inferred from the electromotive forces of its constituents alone without a description, in terms of relational concepts, of the way in which the batteries are connected with each other.[17]

(3) Finally, the predictability of a given characteristic of an object on the basis of specified information concerning its parts will obviously depend on what general laws or theories are available.[18] Thus, the flow of an electric current in a wire connecting a piece of copper and a piece of zinc which are partly immersed in sulfuric acid is unexplainable, on the basis of information concerning any nontrivial set of attributes of copper, zinc, and sulphuric acid, and the particular structure of the system under consideration, unless the theory available contains certain general laws concerning the functioning of batteries, or even more comprehensive principles of physical chemistry. If the theory includes such laws, on the other hand, then the occurrence of the current is predictable. Another illustration, which at the same time provides a good example for the point made under (2) above, is afforded by the optical activity of certain substances. The optical activity of sarco-latic acid, for example, that is, the fact that in solution it rotates the plane of polarization of plane-polarized light, cannot be predicted on the basis of the chemical characteristics of its constituent elements; rather, certain facts about the relations of the atoms constituting a molecule of sarco-lactic acid have to be known. The essential point is that the molecule in question contains an asymmetric carbon atom, that is, one that holds four different atoms or groups, and if this piece of relational information is provided, the optical activity of the solution can be predicted provided that furthermore the theory available for the purpose embodies the law that the presence of one asymmetric carbon atom in a molecule implies optical activity of the solution; if the theory does not include this micro-macro law, then the phenomenon is emergent with respect to that theory.

An argument is sometimes advanced to the effect that phenomena such as the flow of the current, or the optical activity, in our last examples, are absolutely emergent at least in the sense that they could not possibly have been predicted before they had been observed for the first time; in other words, that the laws requisite for their prediction could not have been arrived at on the basis of information available before their first occurrence.[19] This view is untenable, however. On the strength of data available at a given time, science often establishes generalizations by means of which it can forecast the occurrence of events the like of which have never before been encountered. Thus, generalizations based upon periodicities exhibited by the characteristics of chemical elements then known enabled Mendeleev in 1871 to predict the existence of a certain new element and to state correctly various properties of that element as well as of several of its compounds; the element in question, germanium, was not discovered until 1886. A more recent illustration of the same point is provided by the development of the atomic bomb and the prediction, based on theoretical principles established prior to the event, of its explosion under specified conditions, and of its devastating release of energy.

As Grelling has stressed, the observation that the predictability of the occurrence of any characteristic depends upon the theoretical knowledge available, applies even to those cases in which, in the language of some emergentists, the characteristic of the whole is a mere resultant of the corresponding characteristics of the parts and can be obtained from the latter by addition. Thus, even the weight of a water molecule cannot be derived from the weights of its atomic constituents without the aid of a law which expresses the former as some specific mathematical function of the latter. That this function should be the sum is by no means self-evident; it is an empirical generalization, and at that not a strictly correct one, as relativistic physics has shown.

Failure to realize that the question of the predictability of a phenomenon cannot be significantly raised unless the theories available for the prediction have been specified has encouraged the misconception that certain phenomena have a mysterious quality of absolute unexplainability, and that their emergent status has to be accepted with "natural piety," as C. L. Morgan put it. The observations presented in the preceding discussion strip the idea of emergence of these unfounded connotations: emergence of a characteristic is not an ontological trait inherent in some phenomena; rather it is indicative of the scope of our knowledge at a given time; thus it has no absolute, but a relative character; and what is emergent with respect to the theories available today may lose its emergent status tomorrow.

The preceding considerations suggest the following *redefinition of emergence:* The occurrence of a characteristic W in an object w is emergent relative to a theory T, a part relation Pt, and a class G of attributes if that occurrence cannot be deduced by means of T from a characterization of the Pt-parts of w with respect to all the attributes in G.

This formulation explicates the meaning of emergence with respect to *events* of a certain kind, namely the occurrence of some characteristic W in an object w. Frequently, emergence is attributed to *characteristics* rather than to events; this use of the concept of emergence may be interpreted as follows: A characteristic W is emergent relatively to T, Pt, and G if its occurrence in *any* object is emergent in the sense just indicated.

As far as its cognitive content is concerned, the emergentist assertion that the phenomena of life are emergent may now be construed, roughly, as an elliptic formulation of the following statement: Certain specifiable biological phenomena cannot be explained, by means of contemporary physico-chemical theories, on the basis of data concerning the physical and chemical characteristics of the atomic and molecular constituents of organisms. Similarly, the thesis of an emergent status of mind might be taken to assert that present-day physical, chemical, and biological theories do not suffice to explain all psychological phenomena on the basis of data concerning the physical, chemical, and biological characteristics of the cells or of the molecules or atoms constituting the organisms in question. But in this interpretation, the emergent character of biological and psychological phenomena becomes trivial; for the description of various biological phenomena requires terms which are not contained in the vocabulary of present-day physics and chemistry; hence, we cannot expect that all specifically biological phenomena are explainable, that is, deductively inferable, by means of present-day physico-chemical theories on the basis of initial conditions which themselves are described in exclusively physico-chemical terms. In order to obtain a less trivial interpretation of the assertion that the phenomena of life are emergent, we have therefore to include in the explanatory theory all those presumptive laws presently accepted which connect the physico-chemical with the biological "level," that is, which contain, on the one hand, certain physical and chemical terms, including those required for the description of molecular structures, and on the other hand, certain concepts of biology. An analogous observation applies to the case of psychology, If the assertion that life and mind have an emergent status is interpreted in this sense, then its import can be summarized approximately by the statement that no explanation, in terms of micro-structure theories, is available at present for large classes of phenomena studied in biology and psychology.[20]

Assertions of this type, then, appear to represent the rational core of the doctrine of emergence. In its revised form, the idea of emergence no longer carries with it the connotation of absolute unpredictability—a notion which is objectionable not only because it involves and perpetuates certain logical misunderstandings, but also because, not unlike the ideas of neovitalism, it encourages an attitude of resignation which is stifling· for scientific research. No doubt it is this characteristic, together with its theoretical sterility, which accounts for the rejection, by the majority of contemporary scientists, of the classical absolutistic doctrine of emergence.[21]

Part III. Logical Analysis of Law and Explanation

6. Problems of the Concept of General Law

From our general survey of the characteristics of scientific explanation, we now turn to a closer examination of its logical structure. The explanation of a phenomenon, we noted, consists in its subsumption under laws or under a theory. But what is a law, what is a theory? While the meaning of these concepts seems intuitively clear, an attempt to construct adequate explicit definitions for them encounters considerable difficulties. In the present section, some basic problems of the concept of law will be described and analyzed; in the next section, we intend to propose, on the basis of the suggestions thus obtained, definitions of law and of explanation for a formalized model language of a simple logical structure.

The concept of law will be construed here so as to apply to true statements only. The apparently plausible alternative procedure of requiring high confirmation rather than truth of a law seems to be inadequate: It would lead to a relativized concept of law, which would be expressed by the phrase "sentence S is a law relative to the evidence *E*." This does not accord with the meaning customarily assigned to the concept of law in science and in methodological inquiry. Thus, for example, we would not say that Bode's general formula for the distance of the planets from the sun was a law relative to the astronomical evidence available in the 1770s, when Bode propounded it, and that it ceased to be a law after the discovery of Neptune and the determination of its distance from the sun; rather, we would say that the limited original evidence had given a high probability to the assumption that the formula was a law, whereas more recent additional information reduced that probability so much as to make it practically certain that Bode's formula is not generally true, and hence not a law.[22]

Apart from being true, a law will have to satisfy a number of additional conditions. These can be studied independently of the factual requirement of truth, for they refer, as it were, to all logically possible laws, no matter whether factually true or false. Adopting a term proposed by Goodman,[23] we will say that a sentence is *lawlike* if it has all the characteristics of a general law, with the possible exception of truth. Hence, every law is a lawlike sentence, but not conversely.

Our problem of analyzing the notion of law thus reduces to that of explicating the concept of lawlike sentence. We shall construe the class of lawlike sentences as including analytic general statements, such as 'A rose is a rose', as well as the lawlike sentences of empirical science, which have empirical content.[24] It will not be necessary to require that each lawlike sentence permissible in explanatory contexts be of the second kind; rather, our definition of explanation will be so constructed as to guarantee the factual character of the totality of the laws—though not of every single one of them—which function in an explanation of an empirical fact.

What are the characteristics of lawlike sentences? First of all, lawlike sentences are statements of universal form, such as 'All robins' eggs are greenish-blue', 'All metals are conductors of electricity', 'At constant pressure, any gas expands with increasing temperature'. As these examples illustrate, a lawlike sentence usually is not only of universal, but also of conditional form; it makes an assertion to the effect that universally, if a certain set of conditions, *C*, is realized, then another specified set of conditions, *E*, is realized as well. The standard form for the symbolic expression of a lawlike sentence is therefore the universal conditional. However, since any conditional statement can be transformed into a non-conditional one, conditional form will not be considered as essential for a lawlike sentence, while universal character will be held indispensable.

But the requirement of universal form is not sufficient to characterize lawlike sentences. Suppose, for example, that a given basket, *b*, contains at a certain time *t* a number of red apples and nothing else.[25] Then the statement

(S_1) Every apple in basket *b* at time *t* is red

is both true and of universal form. Yet the sentence does not qualify as a law; we would refuse, for example, to explain by subsumption under it the fact that a particular apple chosen at random from the basket is red. What distinguishes S_1 from a lawlike sentence? Two points suggest themselves, which will be considered in turn, namely, finite scope, and reference to a specified object.

First, the sentence S_1 makes in effect an assertion about a finite number of objects only, and this seems irreconcilable with the claim to universality which is commonly associated with the notion of law.[26] But are not Kepler's laws considered as lawlike although they refer to a finite set of planets only? And might we not even be willing to consider as lawlike a sentence such as the following:

(S₂) All the sixteen ice cubes in the freezing tray of this refrigerator have a temperature of less than 10 degrees centigrade.

This point might well be granted; but there is an essential difference between S_1, on the one hand, and Kepler's laws, as well as S_2, on the other: The latter, while finite in scope, are known to be consequences of more comprehensive laws whose scope is not limited, while for S_1 this is not the case.

Adopting a procedure recently suggested by Reichenbach,[27] we will therefore distinguish between fundamental and derivative laws. A statement will be called a derivative law if it is of universal character and follows from some fundamental laws. The concept of fundamental law requires further clarification; so far, we may say that fundamental laws, and similarly fundamental lawlike sentences, should satisfy a certain condition of nonlimitation of scope.

It would be excessive, however, to deny the status of fundamental lawlike sentence to all statements which, in effect, make an assertion about a finite class of objects only, for that would rule out also a sentence such as 'All robins' eggs are greenish-blue', since presumably the class of all robins' eggs—past, present, and future—is finite. But again, there is an essential difference between this sentence and, say, S_1. It requires empirical knowledge to establish the finiteness of the class of robins' eggs, whereas, when the sentence S_1 is construed in a manner which renders it intuitively unlawlike, the terms 'basket b' and 'apple' are understood so as to imply finiteness of the class of apples in the basket at time t. Thus, so to speak, the meaning of its constitutive terms alone—without additional factual information— entails that S_1 has a finite scope. Fundamental laws, then, will have to be construed so as to satisfy a condition of nonlimited scope; our formulation of that condition however, which refers to what is entailed by "the meaning" of certain expressions, is too vague and will have to be revised later. Let us note in passing that the stipulation here envisaged would bar from the class of fundamental lawlike sentences also such undesirable candidates as 'All uranic objects are spherical', where 'uranic' means the property of being the planet Uranus; indeed, while this sentence has universal form, it fails to satisfy the condition of nonlimited scope.

In our search for a general characterization of lawlike sentences, we now turn to a second clue which is provided by the sentence S_1. In addition to violating the condition of nonlimited scope, that sentence has the peculiarity of making reference to a particular object, the basket b; and this, too, seems to violate the universal character of a law.[28] The restriction which seems indicated here, should again be applied to fundamental lawlike sentences only; for a true general statement about the free fall of physical bodies on the moon, while referring to a particular object, would still constitute a law, albeit a derivative one.

It seems reasonable to stipulate, therefore, that a fundamental lawlike sentence must be of universal form and must contain no essential—that is, uneliminable— occurrences of designations for particular objects. But this is not sufficient; indeed, just at this point, a particularly serious difficulty presents itself. Consider the sentence

(S₃) Everything that is either an apple in basket b at time t or a sample of ferric oxide is red.

If we use a special expression, say '*x* is ferple', as synonymous with '*x* is either an apple in *b* at *t* or a sample of ferric oxide', then the content of S_3 can be expressed in the form.

(S_4) Everything that is ferple is red.

The statement thus obtained is of universal form and contains no designations of particular objects, and it also satisfies the condition of nonlimited scope; yet clearly, S_4 can qualify as a fundamental lawlike sentence no more than can S_3.

As long as 'ferple' is a defined term of our language, the difficulty can readily be met by stipulating that after elimination of defined terms, a fundamental lawlike sentence must not contain essential occurrences of designations for particular objects. But this way out is of no avail when 'ferple', or another term of its kind, is a primitive predicate of the language under consideration. This reflection indicates that certain restrictions have to be imposed upon those predicates—that is, terms for properties or relations—which may occur in fundamental lawlike sentences.[29]

More specifically, the idea suggests itself of permitting a predicate in a fundamental lawlike sentence only if it is purely universal, or, as we shall say, purely qualitative, in character; in other words, if a statement of its meaning does not require reference to any one particular object or spatio-temporal location. Thus, the terms 'soft', 'green', 'warmer than', 'as long as', 'liquid', 'electrically charged', 'female', 'father of', are purely qualitative predicates, while 'taller than the Eiffel Tower', 'medieval', 'lunar', 'arctic', 'Ming' are not.[30]

Exclusion from fundamental lawlike sentences of predicates which are not purely qualitative would at the same time ensure satisfaction of the condition of nonlimited scope; for the meaning of a purely qualitative predicate does not require a finite extension; and indeed, all the sentences considered above which violate the condition of nonlimited scope make explicit or implicit reference to specific objects.

The stipulation just proposed suffers, however, from the vagueness of the concept of purely qualitative predicate. The question whether indication of the meaning of a given predicate in English does or does not require reference to some specific object does not always permit of an unequivocal answer, since English as a natural language does not provide explicit definitions or other clear explications of meaning for its terms. It seems therefore reasonable to attempt definition of the concept of law not with respect to English or any other natural language, but rather with respect to a formalized language—let us call it a model language *L*—which is governed by a well-determined system of logical rules, and in which every term either is characterized as primitive or is introduced by an explicit definition in terms of the primitives.

This reference to a well-determined system is customary in logical research and is indeed quite natural in the context of any attempt to develop precise criteria for certain logical distinctions. But it does not by itself suffice to overcome the specific difficulty under discussion. For while it is not readily possible to characterize as not purely qualitative all those among the defined predicates in *L* whose definiens contains an essential occurrence of some individual name, our problem remains open for the primitives of the language, whose meanings are not determined by definitions within the language, but rather by semantic rules of interpretation. For we want to permit the interpretation of the primitives of *L* by means of such attributes as blue, hard, solid, warmer, but not by the properties of being a descendant of Napoleon, or an arctic animal, or a Greek statue; and the difficulty is precisely that of stating rigorous criteria for the distinction between the permissible and the nonpermissible interpretations. Thus the problem of finding an adequate definition for purely qualitative attributes now arises again; namely for the concepts of the meta-language in which the semantical interpretation of the primitives is formulated. We may postpone an encounter with the difficulty by presupposing formalization of the semantical meta-language, the meta-meta-language, and so forth, but somewhere, we will have to stop at a nonformalized meta-language; and for it, a characterization of purely qualitative predicates will be needed and will present much the same problems as nonformalized English, with which we began. The characterization of a purely qualitative predicate as one whose meaning can be made explicit without reference to any one particular object points to the intended meaning but does not explicate it precisely, and the problem of an adequate definition of purely qualitative predicates remains open.

There can be little doubt, however, that there exists a large number of predicates which would be rather generally recognized as purely qualitative in the sense here pointed out, and as permissible in the formulation of fundamental lawlike sentences; some examples have been given above, and the list could be readily enlarged. When we speak of purely qualitative predicates, we shall henceforth have in mind predicates of this kind.

In the following section, a model language L of a rather simple logical structure will be described, whose primitives will be assumed to be qualitative in the sense just indicated. For this language, the concepts of law and explanation will then be defined in a manner which takes into account the general observations set forth in the present section.

7. Definition of Law and Explanation for a Model Language

Concerning the syntax of our model language L, we make the following assumptions: L has the syntactical structure of the lower functional calculus without the identity sign. In addition to the signs of negation alternation (disjunction), conjunction, and implication (conditional), and the symbols of universal and existential quantification with respect to individual variables, the vocabulary of L contains individual constants ($'a'$, $'b'$, ...), individual variables ($'x'$, $'y'$, ...), and predicates of any desired finite degree. The latter may include, in particular, predicates of degree 1 ($'P'$, $'Q'$, ...), which express properties of individuals, and predicates of degree 2 ($'R'$, $'S'$, ...), which express dyadic relations among individuals.

For simplicity, we assume that all predicates are primitive, that is, undefined in L, or else that before the criteria subsequently to be developed are applied to a sentence, all defined predicates which it contains are eliminated in favor of primitives.

The syntactical rules for the formation of sentences and for logical inference in L are those of the lower functional calculus. No sentence may contain free variables, so that generality is always expressed by universal quantification.

For later reference, we now define, in purely syntactical terms, a number of auxiliary concepts. In the following definitions, S is always understood to be a sentence in L.

(7.1a)   *S* is formally true (formally false) in *L* if *S* (the denial of *S*) can be proved in *L*, that is, by means of the formal rules of logical inference for *L*. If two sentences are mutually derivable from each other in *L*, they will be called equivalent.

(7.1b)   *S* is said to be a singular, or alternatively, a molecular sentence if *S* contains no variables. A singular sentence which contains no statement connectives is also called atomic. Illustrations: The sentences '*R(a,b)* ⊃ [*P(a)* · ~*Q(a)*]', '~*Q(a)*', '*R(a,b)*', '*P(a)*' are all singular, or molecular; the last two are atomic.

(7.1c)   *S* is said to be a generalized sentence if it consists of one or more quantifiers followed by an expression which contains no quantifiers. *S* is said to be of universal form if it is a generalized sentence and all the quantifiers occurring in it are universal. *S* is called purely generalized (purely universal) if *S* is a generalized sentence (is of universal form) and contains no individual constants. *S* is said to be essentially universal if it is of universal form and not equivalent to a singular sentence. *S* is called essentially generalized if it is generalized and not equivalent to a singular sentence.

Illustrations: '*(x)[P(x)* ⊃ *Q(x)]*', '*(x)R(a,x)*', '*(x)[P(x)* v *P(a)]*', '*(x)[P(x)* v ~*P(x)]*', '*(Ex)[P(x)* · ~*Q(x)]*', '*(Ex)(y)[R(a,x)* · *S(a,y)*]*'],

are all generalized sentences; the first four are of universal form, the first and fourth are purely universal; the first and second are essentially universal, the third being equivalent to the singular sentence '*P(a)*', and the fourth to '*P(a)*v~*P(a)*'. All sentences except the third and fourth are essentially generalized.

Concerning the semantical interpretation of *L*, we lay down the following two stipulations:

(7.2a)   The primitive predicates of *L* are all purely qualitative.

(7.2b)   The universe of discourse of *L*, that is, the domain of objects covered by the quantifiers, consists of all physical objects or of all spatio-temporal locations.

A linguistic framework of the kind here characterized is not sufficient for the formulation of scientific theories, since it contains no functors and does not provide the means for dealing with real numbers. Besides, the question is open at present whether a constitution system can be constructed in which all of the concepts of empirical science are reduced, by chains of explicit definitions, to a basis of primitives of a purely qualitative character. Nevertheless, we consider it worthwhile to study the problems at hand for the simplified type of language just described because the analysis of law and explanation is far from trivial even for our model language *L*, and because that analysis sheds light on the logical character of the concepts under investigation also in their application to more complex contexts.

In accordance with the considerations developed in section 6, we now define:

(7.3a)   *S* is a fundamental lawlike sentence in *L* if *S* is purely universal; *S* is a fundamental law in *L* if *S* is purely universal and true.

(7.3b)   *S* is a derivative law in *L* if (1) *S* is essentially, but not purely, universal and (2) there exists a set of fundamental laws in *L* which has *S* as a consequence.

(7.3c)   *S* is a law in *L* if it is a fundamental or a derivative law in *L*.

The fundamental laws as here defined obviously include, besides general statements of empirical character, all those statements of purely universal form which are true on purely logical grounds; that is, those which are formally true in L, such as '$(x)[P(x)\text{v}\sim P(x)]$', and those whose truth derives exclusively from the interpretation given to its constituents, as is the case with

$$'(x)[P(x) \supset Q(x)]',$$

if 'P' is interpreted as meaning the property of being a father, and 'Q' that of being male. The derivative laws, on the other hand, include neither of these categories; indeed, no fundamental law is also a derivative one.[31]

As the primitives of L are purely qualitative, all the statements of universal form in L also satisfy the requirement of nonlimited scope, and thus it is readily seen that the concept of law as defined above satisfies all the conditions suggested in section 6.[32]

The explanation of a phenomenon may involve generalized sentences which are not of universal form. We shall use the term 'theory' to refer to such sentences, and we define this term by the following chain of definitions:

(7.4a) S is a fundamental theory if S is purely generalized and true.

(7.4b) S is a derivative theory in L if (1) S is essentially, but not purely, generalized and (2) there exists a set of fundamental theories in L which has S as a consequence.

(7.4c) S is a theory in L if it is a fundamental or a derivative theory in L.

By virtue of the above definitions, every law is also a theory, and every theory is true.

With the help of the concepts thus defined, we will now reformulate more precisely our earlier characterization of scientific explanation with specific reference to our model language L. It will be convenient to state our criteria for a sound explanation in the form of a definition for the expression "the ordered couple of sentences, (T,C), constitutes an explanans for the sentence E." Our analysis will be restricted to the explanation of particular events, that is, to the case where the explanandum, E, is a singular sentence.[33]

In analogy to the concept of lawlike sentence, which need not satisfy a requirement of truth, we will first introduce an auxiliary concept of potential explanans, which is not subject to a requirement of truth; the notion of explanans will then be defined with the help of this auxiliary concept.—The considerations presented in Part I suggest the following initial stipulations:

(7.5) An ordered couple of sentences, (T,C), constitutes a potential explanans for a singular sentence E only if
  (1) T is essentially generalized and C is singular
  (2) E is derivable in L from T and C jointly, but not from C alone.

(7.6) An ordered couple of sentences, (T,C), constitutes an explanans for a singular sentence E if and only if
  (1) (T,C) is a potential explanans for E
  (2) T is a theory and C is true.

(7.6) is an explicit definition of explanation in terms of the concept of potential explanation.[34] On the other hand, (7.5) is not suggested as a definition, but as a statement of necessary conditions of potential explanation. These conditions will presently be shown not to be sufficient, and additional requirements will be discussed by which (7.5) has to be supplemented in order to provide a definition of potential explanation.

Before we turn to this point, some remarks are called for concerning the formulation of (7.5). The analysis presented in Part I suggests that an explanans for a singular sentence consists of a class of generalized sentences and a class of singular ones. In (7.5), the elements of each of these classes separately are assumed to be conjoined to one sentence. This provision will simplify our formulations, and in the case of generalized sentences, it serves an additional purpose: A class of essentially generalized sentences may be equivalent to a singular sentence; thus, the class $\{$'$P(a)$v$(x)Q(x)$', '$P(a)$v$\sim(x)Q(x)$'$\}$ is equivalent with the sentence '$P(a)$'. Since scientific explanation makes essential use of generalized sentences, sets of laws of this kind have to be ruled out; this is achieved above by combining all the generalized sentences in the explanans into one conjunction, T, and stipulating that T has to be essentially generalized. Again, since scientific explanation makes essential use of generalized sentences E must not be a consequence of C alone: The law of gravitation, combined with the singular sentence 'Mary is blonde and blue-eyed' does not constitute an explanans for 'Mary is blonde'. The last stipulation in (7.5) introduces the requisite restriction and thus prohibits complete self-explanation of the explanandum, that is, the derivation of E from some singular sentence which has E as a consequence. The same restriction also dispenses with the need for a special requirement to the effect that T has to have factual content if (T,C) is to be a potential explanans for an empirical sentence E. For if E is factual, then, since E is a consequence of T and C jointly, but not of C alone, T must be factual, too.

Our stipulations in (7.5) do not preclude, however, what might be termed partial self-explanation of the explanandum. Consider the sentences $T_1 = $ '$(x)[P(x) \supset Q(x)]$', $C_1 = $ '$R(a,b) \cdot P(a)$', $E_1 = $ '$Q(a) \cdot R(a,b)$'. They satisfy all the requirements laid down in (7.5), but it seems counterintuitive to say that (T_1,C_1) potentially explains E_1, because the occurrence of the component '$R(a,b)$' of C_1 in the sentence E_1 amounts to a partial explanation of the explanandum by itself. Is it not possible to rule out, by an additional stipulation, all those cases in which E shares part of its content with C, that is, where C and E have a common consequence which is not formally true in L? This stipulation would be tantamount to the requirement that C and E have to be exhaustive alternatives in the sense that their disjunction is formally true, for the content which any two sentences have in common is expressed by their disjunction. The proposed restriction, however, would be very severe. For if E does not share even part of its content with C, then C is altogether unnecessary for the derivation of E from T and C, that is, E can be inferred from T alone. Therefore, in every potential explanation in which the singular component of the explanans is not dispensable, the explanandum is partly explained by itself. Take, for example the potential explanation of $E_2 = $ '$Q(a)$' by $T_2 = $ '$(x)[P(x) \supset Q(x)]$' and $C_2 = $ '$P(a)$', which satisfies (7.5), and which surely is intuitively unobjectionable. Its three components may be equivalently expressed by the following sentences:

$$T'_2 = \text{'}(x)[\sim P(x) \text{v} Q(x)]\text{'}; \; C'_2 = \text{'}[P(a) \text{v} Q(a)] \cdot [P(a) \text{v} \sim Q(a)]\text{'};$$
$$E'_2 = \text{'}[P(a) \text{v} Q(a)] \cdot [\sim P(a) \text{v} Q(a)].\text{'}$$

This reformulation shows that part of the content of the explanandum is contained in the content of the singular component of the explanans and is, in this sense, explained by itself.

Our analysis has reached a point here where the customary intuitive idea of explanation becomes too vague to provide further guidance for rational reconstruction. Indeed, the last illustration strongly suggests that there may be no sharp boundary line which separates the intuitively permissible from the counterintuitive types of partial self-explanation; for even the potential explanation just considered, which is acceptable in its original formulation, might be judged unacceptable on intuitive grounds when transformed into the equivalent version given above.

The point illustrated by the last example is stated more explicitly in the following theorem, which we formulate here without proof.

(7.7) *Theorem.* Let (T,C) be a potential explanans for the singular sentence E. Then there exist three singular sentences, E_1, E_2 and C_1 in L such that E is equivalent to the conjunction $E_1 \cdot E_2$, C is equivalent to the conjunction $C_1 \cdot E_1$, and E_2 can be derived in L from T alone.[35]

In more intuitive terms, this means that if we represent the deductive structure of the given potential explanation by the schema $\{T,C\} \to E$, then this schema can be restated in the form $\{T,C_1 \cdot E_1\} \to E_1 \cdot E_2$ where E_2 follows from T alone, so that C_1 is entirely unnecessary as a premise; hence, the deductive schema under consideration can be reduced to $\{T,E_1\} \to E_1 \cdot E_2$, which can be decomposed into the two deductive schemata $\{T\} \to E_2$ and $\{E_1\} \to E_1$. The former of these might be called a purely theoretical explanation of E_2 by T, the latter a complete self-explanation of E_1. Theorem (7.7) shows, in other words, that every explanation whose explanandum is a singular sentence can be decomposed into a purely theoretical explanation and a complete self-explanation; and any explanation of this kind in which the singular constituent of the explanans is not completely unnecessary involves a partial self-explanation of the explanandum.[36]

To prohibit partial self-explanation altogether would therefore mean limiting explanation to purely theoretical explanation. This measure seems too severely restrictive. On the other hand, an attempt to delimit, by some special rule, the permissible degree of self-explanation does not appear to be warranted because, as we saw, customary usage provides no guidance for such a delimitation, and because no systematic advantage seems to be gained by drawing some arbitrary dividing line. For these reasons, we refrain from introducing stipulations to prohibit partial self-explanation.

The conditions laid down in (7.5) fail to preclude yet another unacceptable type of explanatory argument, which is closely related to complete self-explanation, and which will have to be ruled out by an additional stipulation. The point is, briefly, that if we were to accept (7.5) as a definition, rather than merely as a statement of necessary conditions, for potential explanation then, as a consequence of (7.6), any given particular fact could be explained by means of any true lawlike sentence whatsoever. More explicitly, if E is a true sentence—say, 'Mt. Everest is snowcapped', and T is a law—say, 'All metals are good conductors of heat', then there always exists a true singular sentence C such that E is derivable from T and C, but not from C alone; in other words, such that (7.5) is satisfied. Indeed, let T_s be some arbitrarily chosen particular instance of T, such as 'If the Eiffel Tower is metal, it is a good conductor of heat'. Now since E is true, so is the conditional $T_s \supset E$, and if the latter is chosen as the sentence C, then T, C, E satisfy the conditions laid down in (7.5).

In order to isolate the distinctive characteristic of this specious type of explanation, let us examine an especially simple case of the objectionable kind. Let $T_1 =$ '$(x)P(x)$' and $E_1 =$ '$R(a,b)$'; then the sentence $C_1 =$ '$P(a) \supset R(a,b)$' is formed in accordance with the preceding instructions, and T_1, C_1, E_1 satisfy the conditions (7.5). Yet, as the preceding example illustrates, we would not say that (T_1, C_1) constitutes a potential explanans for E_1. The rationale for the verdict may be stated as follows: If the theory T_1 on which the explanation rests, is actually true, then the sentence C_1, which can also be put into the form $\sim P(a) \lor R(a,b)$', can be verified, or shown to be true, only by verifying '$R(a,b)$', that is, E_1. In this broader sense, E_1 is here explained by itself. And indeed, the peculiarity just pointed out clearly deprives the proposed potential explanation for E_1 of the predictive import which, as was noted in Part I, is essential for scientific explanation: E_1 could not possibly be predicted on the basis of T_1 and C_1 since the truth of C_1 cannot be ascertained in any manner which does not include verification of E_1. (7.5) should therefore be supplemented by a stipulation to the effect that if (T,C) is to be a potential explanans for E, then the assumption that T is true must not imply that verification of C necessitates verification of E.[37]

How can this idea be stated more precisely? Study of an illustration will suggest a definition of verification for molecular sentences. The sentence $M =$ '$[\sim P(a) \cdot Q(a)] \lor R(a,b)$' may be verified in two different ways, either by ascertaining the truth of the two sentences '$\sim P(a)$' and '$Q(a)$', which jointly have M as a consequence, or by establishing the truth of the sentence '$R(a,b)$', which again, has M as a consequence. Let us say that S is a basic sentence in L if S is either an atomic sentence or the negation of an atomic sentence in L. Verification of a molecular sentence S may then be defined generally as the establishment of the truth of some class of basic sentences which has S as a consequence. Hence, the intended additional stipulation may be restated: The assumption that T is true must not imply that every class of true basic sentences which has C as a consequence also has E as a consequence.

As brief reflection shows, this stipulation may be expressed in the following form, which avoids reference to truth: T must be compatible in L with at least one class of basic sentences which has C but not E as a consequence; or, equivalently: There must exist at least one class of basic sentences which has C, but neither $\sim T$ nor E as a consequence.

As brief reflection shows, this stipulation may be expressed in the following form, which avoids reference to truth: T must be compatible in L with at least one class of basic sentences which has C but not E as a consequence; or, equivalently: There must exist at least one class of basic sentences which has C, but neither $\sim T$ nor E as a consequence in L.

If this requirement is met, then surely E cannot be a consequence of C, for otherwise there could be no class of basic sentences which has C but not E as a consequence; hence, supplementation of (7.5) by the new condition renders the second stipulation in (7.5) (2) superfluous. We now define potential explanation as follows:

(7.8)   An ordered couple of sentences (T,C), constitutes a potential explanans for a singular sentence if and only if the following conditions are satisfied:
  (1) T is essentially generalized and C is singular
  (2) E is derivable in L from T and C jointly
  (3) T is compatible with at least one class of basic sentences which has C but not E as a consequence.

The definition of the concept of explanans by means of that of potential explanans as formulated in (7.6) remains unchanged.

In terms of our concept of explanans, we can give the following interpretation to the frequently used phrase "this fact is explainable by means of that theory":

(7.9)   A singular sentence E is explainable by a theory T if there exists a singular sentence C such that (T,C) constitutes an explanans for E.

The concept of causal explanation, which has been examined here, is capable of various generalizations. One of these consists in permitting T to include statistical laws. This requires, however, a previous strengthening of the means of expression available in L, or the use of a complex theoretical apparatus in the meta-language. On the other hand, and independently of the admission of statistical laws among the explanatory principles, we might replace the strictly deductive requirement that E has to be a consequence of T and C jointly by the more liberal inductive one that E has to have a high degree of confirmation relatively to the conjunction of T and C. Both of these extensions of the concept of explanation open important prospects and raise a variety of new problems. In the present essay, however, these issues will not be further pursued.

Part IV. The Systematic Power of a Theory

8. Explication of the Concept of Systematic Power

Scientific laws and theories have the function of establishing systematic connections among the data of our experience, so as to make possible the derivation of some of those data from others. According as, at the time of the derivation, the derived data are, or are not yet, known to have occurred, the derivation is referred to as explanation or as prediction. Now it seems sometimes possible to compare different theories, at least in an intuitive manner, in regard to their explanatory or predictive powers: Some theories seem powerful in the sense of permitting the derivation of many data from a small amount of initial information; others seem less powerful, demanding comparatively more initial data, or yielding fewer results. Is it possible to give a precise interpretation to comparisons of this kind by defining, in a completely general manner, a numerical measure for the explanatory or predictive power of a theory? In the present section, we shall develop such a definition and examine some of its implications; in the following section, the definition will be expanded and a general theory of the concept under consideration will be outlined.

Since explanation and prediction have the same logical structure, namely that of a deductive systematization, we shall use the neutral term "systematic power" to refer to the intended concept. As is suggested by the preceding intuitive characterization, the systematic power of a theory T will be reflected in the ratio of the amount of information derivable by means of T to the amount of initial information required for that derivation. This ratio will obviously depend on the particular set of data, or of information, to which T is applied, and we shall therefore relativize our concept accordingly. Our aim, then, is to construct a definition for $s(T,K)$, the systematic power of a theory T with respect to a finite class K of data, or the degree to which T deductively systematizes the information contained in K.

Our concepts will be constructed again with specific reference to the language L. Any singular sentence in L will be said to express a potential datum, and K will accordingly be construed as a finite class of singular sentences.[38] T will be construed in a much broader sense than in the preceding sections; it may be any sentence in L, no matter whether essentially generalized or not. This liberal convention is adopted in the interest of the generality and simplicity of the definitions and theorems now to be developed.

To obtain values between 0 and 1 inclusive, we might now try to identify $s(T,K)$ with the percentage of those sentences in K which are derivable from the remainder by means of T. Thus, if $K_1 = \{$ '$P(a)$', '$Q(a)$', '$\sim P(b)$', '$\sim Q(b)$', '$Q(c)$', '$\sim P(d)$' $\}$, and $T_1 = $ '$(x)[P(x) \supset Q(x)]$', then exactly the second and third sentence in K_1 are derivable by means of T_1 from the remainder, in fact from the first and fourth sentence. We might therefore consider setting $s(T_1,K_1) = 2/6 = 1/3$. But then, for the class $K_2 = \{$ '$P(a) \cdot Q(a)$', '$\sim P(b) \cdot \sim Q(b)$', '$Q(c)$', '$\sim P(d)$' $\}$, the same T_1 would have the s-value 0, although K_2 contains exactly the same information as K_1; again for yet another formulation of that information, namely, $K_3 = \{$ '$P(a) \cdot \sim Q(b)$', '$Q(a) \cdot \sim P(b)$', '$Q(c)$', '$\sim P(d)$' $\}$, T_1 would have the s-value $1/4$, and so on. But what we seek is a measure of the degree to which a given theory deductively systematizes a given body of factual information, that is, a certain content, irrespective of the particular structure and grouping of the sentences in which that content happens to be expressed. We shall therefore make use of a method which represents the content of any singular sentence or class of singular sentences as composed of certain uniquely determined smallest bits of information. By applying our general idea to these bits, we shall obtain a measure for the systematic power of T and K which is independent of the way in which the content of K is formulated. The sentences expressing those smallest bits of information will be called minimal sentences and an exact formulation of the proposed procedure will be made possible by an explicit definition of this auxiliary concept. To this point we now turn.

If, as will be assumed here, the vocabulary of L contains fixed finite numbers of individual constants and of predicate constants, then only a certain finite number, say n, of different atomic sentences can be formulated in L. By a minimal sentence in L, we will understand a disjunction of any number $k(0 \leq k \leq n)$ of different atomic sentences and the negations of the $n - k$ remaining ones. Clearly, n atomic sentences determine 2^n minimal sentences. Thus, if a language L_1 contains exactly one individual constant, 'a', and exactly two primitive predicates, 'P' and 'Q', both of degree 1, then L_1 contains two atomic sentences, '$P(a)$' and '$Q(a)$', and four minimal sentences, namely, '$P(a) \lor Q(a)$', '$P(a) \lor \sim Q(a)$', '$\sim P(a) \lor Q(a)$', '$\sim P(a) \lor \sim Q(a)$'. If another language, L_2, contains in addition to the vocabulary of L_1 a second individual constant, 'b', and a predicate 'R' of degree 2, then L_2 contains eight atomic sentences and 256 minimal sentences, such as '$P(a) \lor P(b) \lor \sim Q(a) \lor Q(b) \lor R(a,a) \lor R(a,b) \lor \sim R(b,a) \lor \sim R(b,b)$'.

The term 'minimal sentence' is to indicate that the statements in question are the singular sentences of smallest non-zero content in L, which means that every singular sentence in L which follows from a minimal sentence is either equivalent to that minimal sentence or logically true in L. Minimal sentences do have consequences other than themselves which are not logically true in L, but these are not of singular form; '$(Ex)(P(x)vQ(x))$' is such a consequence of '$P(a)vQ(a)$' in L_1 above.

Furthermore, no two minimal sentences have any consequence in common which is not logically, or formally, true in L; in other words, the contents of any two minimal sentences are mutually exclusive.

By virtue of the principles of the sentential calculus, every singular sentence which is not formally true in L can be transformed into a conjunction of uniquely determined minimal sentences; this conjunction will be called the minimal normal form of the sentence. Thus, for example, in the language L_1 referred to above, the sentences '$P(a)$' and '$Q(a)$' have the minimal normal forms '$[P(a)vQ(a)] \cdot [P(a)v \sim Q(a)]$', and '$[P(a)vQ(a)] \cdot [\sim P(a)vQ(a)]$', respectively; in L_2, the same sentences have minimal normal forms consisting of 128 conjoined minimal sentences each. If a sentence is formally true in L, its content is zero, and it cannot be represented by a conjunction of minimal sentences. It will be convenient, however, to say that the minimal normal form of a formally true sentence in L is the vacuous conjunction of minimal sentences, which does not contain a single term.

As a consequence of the principle just mentioned, any class of singular sentences which are not all formally true can be represented by a sentence in minimal normal form. The basic idea outlined above for the explication of the concept of systematic power can now be expressed by the following definition:

(8.1) Let T be any sentence in L, and K any finite class of singular sentences in L which are not all formally true. If K' is the class of minimal sentences which occur in the minimal normal form of K, consider all divisions of K' into two mutually exclusive subclasses, K_1' and K_2', such that every sentence in K_2' is derivable from K_1' by means of T. Each division of this kind determines a ratio $n(K_2')/n(K')$, that is, the number of minimal sentences in K_2' divided by the total number of minimal sentences in K'. Among the values of these ratios, there must be a largest one; $s(T, K)$ is equal to that maximum ratio. (Note that if all the elements of K were formally true, $n(K')$ would be 0 and the above ratio would not be defined.)

Illustration: Let L_1 contain only one individual constant, 'a', and only two predicates, 'P' and 'Q', both of degree 1. In L_1, let $T = $ '$(x)[P(x) \supset Q(x)]$', $K = $ \{'$P(a)$', '$Q(a)$'\}. Then we have $K' = $ \{'$P(a)vQ(a)$', '$P(a)v \sim Q(a)$', '$\sim P(a)vQ(a)$'\}. From the subclass K_1' consisting of the first two elements of K'—which together are equivalent to '$P(a)$'—we can derive, by means of T, the sentence '$Q(a)$', and from it, by pure logic, the third element of K'; it constitutes the only element of K_2'. No "better" systematization is possible, hence, $s(T, K) = 1/3$.

Our definition leaves open, and is independent of, the question whether for a given K' there might not exist different divisions each of which would yield the maximum value for $n(K_2')/n(K')$. Actually, this can never happen: there exists always exactly one optimal subdivision of a given K'. This fact is a corollary of a general theorem, to which we now turn. It will be noticed that in the last illustration, K_2' can be derived from T alone, without the use of K_1' as a premise; indeed, '$\sim P(a)vQ(a)$' is but a substitution instance of the sentence '$(x)[\sim P(x)vQ(x)]$', which is equivalent to T. The theorem now to be formulated, which might appear surprising at first, shows that this observation applies analogously in all other cases.

(8.2) *Theorem.* Let T be any sentence, K' a class of minimal sentences, and K_2' a subclass of K' such that every sentence in K_2' is derivable by means of T from the class $K—K_2'$; then every sentence in K_2' is derivable from T alone.

The proof, in outline, is as follows: Since the contents of any two different minimal sentences are mutually exclusive, so must be the contents of K_1' and K_2', which have not a single minimal sentence in common. But since the sentences of K_2' follow from K_1' and T jointly, they must therefore follow from T alone.

We note the following consequences of our theorem:

(8.2a) *Theorem.* In any class K' of minimal sentences, the largest subclass which is derivable from the remainder by means of a sentence T is identical with the class of those elements in K' which are derivable from T alone.

(8.2b) *Theorem.* Let T be any sentence, K a class of singular sentences which are not all formally true, K' the equivalent class of minimal sentences, and K_1' the class of those among the latter which are derivable from T alone. Then the concept s defined in (8.1) satisfies the following equation:

$$s(T, K) = n(K_t')/n(K')$$

9. *Systematic Power and Logical Probability of a Theory.*
Generalization of the Concept of Systematic Power

The concept of systematic power is closely related to that of degree of confirmation, or logical probability, of a theory. A study of this relationship will shed new light on the proposed definition of s, will suggest certain ways of generalizing it, and will finally lead to a general theory of systematic power which is formally analogous to that of logical probability.

The concept of logical probability, or degree of confirmation, is the central concept of inductive logic. Recently, different explicit definitions for this concept have been proposed, for languages of a structure similar to that of our model language, by Carnap[39] and by Helmer, Hempel, and Oppenheim.[40]

While the definition of s proposed in the preceding section rests on the concept of minimal sentence, the basic concept in the construction of a measure for logical probability is that of state description or, as we shall also say, of maximal sentence. A maximal sentence is the dual[41] of a minimal sentence in L; it is a conjunction of $k(0 \leq k \leq n)$ different atomic sentences and of the negations of the remaining $n - k$ atomic sentences. In a language with n atomic sentences, there exist 2^n state descriptions. Thus, for example, the language L_1 repeatedly mentioned in §8 contains the following four maximal sentences: '$P(a) \cdot Q(a)$', '$P(a) \cdot {\sim}Q(a)$', '${\sim}P(a) \cdot Q(a)$', '${\sim}P(a) \cdot {\sim}Q(a)$'.

The term "maximal sentence" is to indicate that the sentences in question are the singular sentences of maximum nonuniversal content in L, which means that every singular sentence in L which has a maximal sentence as a consequence is either equivalent with that maximal sentence or formally false in L.

As we saw, every singular sentence can be represented in a conjunctive, or minimal form, that is, as a conjunction of certain uniquely determined minimal sentences; similarly, every singular sentence can be expressed also in a disjunctive, or maximal, normal form, that is, as a disjunction of certain uniquely determined maximal sentences. In the language L_1, for example, '$P(a)$' has the minimal normal form '$[P(a)vQ(a)] \cdot [P(a)v{\sim}Q(a)]$' and the maximal normal form '$[P(a) \cdot Q(a)]v[P(a) \cdot {\sim}Q(a)]$'; the sentence '$P(a) \supset Q(a)$' has the minimal normal form '${\sim}P(a)vQ(a)$' and the maximal normal form '$[P(a) \cdot Q(a)]v[{\sim}P(a) \cdot Q(a)]v[{\sim}P(a) \cdot {\sim}Q(a)]$'; the minimal normal form of a formally true sentence is the vacuous conjunction, while its maximal normal form is the disjunction of all four state descriptions in L_1. The minimal normal form of any formally false sentence is the conjunction of all four minimal sentences in L_1, while its maximal normal form is the vacuous disjunction, as we shall say.

The minimal normal form of a singular sentence is well suited as an indicator of its content, for it represents the sentence as a conjunction of standard components whose contents are minimal and mutually exclusive. The maximal normal form of a sentence is suited as an indicator of its range, that is, intuitively speaking, of the variety of its different possible realizations, or of the variety of those possible states of the world which, if realized, would make the statement true. Indeed, each maximal sentence may be said to describe, as completely as the means of L permit, one possible state of the world; and the state descriptions constituting the maximal normal form of a given singular sentence simply list those among the possible states which would make the sentence true.

Just like the contents of any two different minimal sentences, the ranges of any two maximal sentences are mutually exclusive: no possible state of the world can make two different maximal sentences true because any two maximal sentences are obviously incompatible with each other.[42]

Range and content of a sentence vary inversely. The more a sentence asserts, the smaller the variety of its possible realizations, and conversely. This relationship is reflected in the fact that the larger number of constituents in the minimal normal form of a singular sentence, the smaller the number of constituents in its maximal normal form, and conversely. In fact, if the minimal normal form of a singular sentence U contains m_U of the $m = 2^n$ minimal sentences in L, then its maximal normal form contains $I_U = m - m_U$ of the m maximal sentences in L. This is illustrated by our last four examples, where $m = 4$, and $m_U = 2, 1, 0, 4$ respectively.

The preceding observations suggest that the content of any singular sentence U might be measured by the corresponding number m_U or by some magnitude proportional to it. Now it will prove convenient to restrict the values of the content measure function to the interval from 0 to 1, inclusive; and therefore, we define a measure, $g_1(U)$, for the content of any singular sentence in L by the formula:

(9.1) $$g_1(U) = m_U/m$$

To any finite class K of singular sentences, we assign, as a measure $g_1(K)$ of its content, the value $g_1(S)$, where S is the conjunction of the elements of K.

By virtue of this definition, the equation in theorem (8.2b) may be rewritten:

$$s(T, K) = g_1(K'_t)/g_1(K')$$

Here, K'_t is the class of all those minimal sentences in K' which are consequences of T. In the special case where T is a singular sentence, K' is therefore equivalent with TvS, where S is the conjunction of all the elements of K'. Hence, the preceding equation may then be transformed into

(9.2) $$s(T, S) = g_1(TvS)/g_1(S)$$

This formula holds when T and S are singular sentences, and S is not formally true. It bears a striking resemblance to the general schema for the definition of the logical probability of T in regard to S:

(9.3) $$p(T, S) = r(T \cdot S)/r(S)$$

Here $r(U)$ is, for any sentence U in L, a measure of the range of U, T is any sentence in L, and S any sentence in L with $r(S) \neq 0$.

The several specific definitions which have been proposed for the concept of logical probability accord essentially with the pattern exhibited by (9.3),[43] but they differ in their choice of a specific measure function for ranges, that is, in their definition of r. One idea which comes to mind is to assign, to any singular sentence U whose maximal normal form contains l_U maximal sentences, the range measure

(9.4) $$r_1(U) = l_U/m$$

which obviously is defined in strict analogy to the content measure g_1 for singular sentences as introduced in (9.1). For every singular sentence U, the two measures add up to unity:

(9.5) $$r_1(U) + g_1(U) = (l_U + m_U)/m = 1$$

As Carnap has shown, however, the range measure r_1 confers upon the corresponding concept of logical probability, that is, upon the concept p_1 defined by means of it according to the schema (9.3), certain characteristics which are incompatible with the intended meaning of logical probability;[44] and Carnap, as well as Helmer jointly with the present authors, has suggested certain alternative measure functions for ranges, which lead to more satisfactory concepts of probability or of degree of confirmation. While we need not enter into details here, the following general remarks seem indicated to prepare the subsequent discussion.

The function r_1 measures the range of a singular sentence essentially by counting the number of maximal sentences in its maximal normal form; it thus gives equal weight to all maximal sentences (definition (9.1) deals analogously with minimal sentences). The alternative definitions just referred to are based on a different procedure. Carnap, in particular, lays down a rule which assigns a specific weight, that is, a specific value of r, to each maximal sentence, but these weights are not the same for all maximal sentences. He then defines the range measure of any other singular sentence as the sum of the measures of its constituent maximal sentences. In terms of the function thus obtained—let us call it r_2—Carnap defines the corresponding concept of logical probability, which we shall call p_2, for singular sentences

T, S in accordance with the schema (9.3): $p_2(T, S) = r_2(T \cdot S)/r_2(S)$. The definitions of r_2 and p_2 are then extended, by means of certain limiting processes, to the cases where T and S are no longer both singular.[45]

Now it can readily be seen that just as the function r_1 defined in (9.5) is but one among an infinity of possible range measures, so the analogous function g_1 defined in (9.1) is but one among an infinity of possible content measures; and just as each range measure may serve to define, according to the schema (9.3), a corresponding measure of logical probability, so each content measure function may serve to define, by means of the schema illustrated by (9.2), a corresponding measure of systematic power. The method which suggests itself here for obtaining alternative content measure functions is to choose some suitable range measure r other than r_1 and then to define a corresponding content measure g in terms of it by means of the formula

(9.6) $$g(U) = 1 - r(U)$$

so that g and r satisfy the analogue to (9.5) by definition. The function g thus defined will lead in turn, via a definition analogous to (9.2), to a corresponding concept s. Let us now consider this procedure a little more closely.

We assume that a function r is given which satisfies the customary requirements for range measures, namely:

(9.7) 1. $r(U)$ is uniquely determined for all sentences U in L.
      2. $0 \leqq r(U) \leqq 1$ for every sentence U in L.
      3. $r(U) = 1$ if the sentence U is formally true in L and thus has universal range.
      4. $r(U_1 \vee U_2) = r(U_1) + r(U_2)$ for any two sentences U_1, U_2 whose ranges are mutually exclusive, that is, whose conjunction is formally false.

In terms of the given range measure let the corresponding content measure g be defined by means of (9.6). Then g can readily be shown to satisfy the following conditions:

(9.8) 1. $g(U)$ is uniquely determined for all sentences U in L.
      2. $0 \leqq g(U) \leqq 1$ for every sentence U in L.
      3. $g(U) = 1$ if the sentence U is formally false in L and thus has universal content.
      4. $g(U_1 \cdot U_2) = g(U_1) + g(U_2)$ for any two sentences U_1, U_2 whose contents are mutually exclusive, that is, whose disjunction is formally true.

In analogy to (9.2), we shall next define, by means of g, a corresponding function s:

(9.9) $$s(T,S) = g(T \vee S)/g(S)$$

This function is determined for every sentence T, and for every sentence S with $g(S) \neq 0$, whereas the definition of systematic power given in §8 was restricted to those cases where S is singular and not formally true. Finally, our range measure r determines a corresponding probability function by virtue of the definition

(9.10) $$p(T,S) = r(T \cdot S)/r(S)$$

This formula determines the function p for any sentence T, and for any sentence S with $r(S) \neq 0$.

In this manner, every range measure r which satisfies (9.7) determines uniquely a corresponding content measure g which satisfies (9.8), a corresponding function s, defined by (9.9), and a corresponding function p, defined by (9.10). As a consequence of (9.7) and (9.10), the function p can be shown to satisfy the elementary laws of probability theory, especially those listed in (9.12) below; and by virtue of these, it is finally possible to establish a very simple relationship which obtains, for any given range measure r, between the corresponding concepts $p(T,S)$ and $s(T,S)$. Indeed, we have

(9.11)
$$
\begin{aligned}
s(T,S) &= g(T v S)/g(S) \\
&= [1 - r(T v S)]/[1 - r(S)] \\
&= r[\sim(T v S)]/r(\sim S) \\
&= r(\sim T \cdot \sim S)/r(\sim S) \\
&= p(\sim T, \sim S)
\end{aligned}
$$

We now list, without proof, some theorems concerning p and s which follow from our assumptions and definitions; they hold in all cases where the values of p and s referred to exist, that is, where the r-value of the second argument of p, and the g-value of the second arguments of s, is not 0.

(9.12) (1) a. $\qquad\qquad 0 \leq p(T,S) \leq 1$
      b. $\qquad\qquad 0 \leq s(T,S) \leq 1$
    (2) a. $\qquad p(\sim T,S) = 1 - p(T,S)$
      b. $\qquad s(\sim T,S) = 1 - s(T,S)$
    (3) a. $p(T_1 v T_2, S) = p(T_1,S) + p(T_2,S) - p(T_1 \cdot T_2, S)$
      b. $s(T_1 \cdot T_2, S) = s(T_1,S) + s(T_2,S) - s(T_1 v T_2, S)$
    (4) a. $p(T_1 \cdot T_2, S) = p(T_1,S) \cdot p(T_2, T_1 \cdot S)$
      b. $s(T_1 v T_2, S) = s(T_1,S) \cdot s(T_2, T_1 v S)$

In the above grouping, these theorems exemplify the relationship of dual correspondence which obtains between p and s. A general characterization of this correspondence is given in the following theorem, which can be proved on the basis of (9.11), and which is stated here in a slightly informal manner in order to avoid the tedium of lengthy formulations.

(9.13) *Dualism theorem.* From any demonstrable general formula expressing an equality or an inequality concerning p, a demonstrable formula concerning s is obtained if 'p' is replaced, throughout, by 's', and '$\cdot$' and 'v' are exchanged for each other. The same exchange, and replacement of 's' by 'p', conversely transforms any theorem expressing an equality or an inequality concerning s into a theorem about p.

We began our analysis of the systematic power of a theory in regard to a class of data by interpreting this concept, in §8, as a measure of the optimum ratio of those among the given data which are derivable from the remainder by means of the theory. Systematic elaboration of this idea has led to the definition, in the present section, of a more general concept of systematic power, which proved to be the dual counterpart of the concept of logical probability. This extension of our original interpretation yields a simpler and more comprehensive theory than would have been attainable on the basis of our initial definition.

But the theory of systematic power, in its narrower as well as in its generalized version, is, just like the theory of logical probability, purely formal in character, and a significant application of either theory in epistemology or the methodology of science requires the solution of certain fundamental problems which concern the logical structure of the language of science and the interpretation of its concepts. One urgent desideratum here is the further elucidation of the requirement of purely qualitative primitives in the language of science; another crucial problem is that of choosing, among an infinity of formal possibilities, an adequate range measure r. The complexity and difficulty of the issues which arise in these contexts has been brought to light by recent investigations;[46] it can only be hoped that recent advances in formal theory will soon be followed by progress in solving those open problems and thus clarifying the conditions for a sound application of the theories of logical probability and of systematic power.

Notes

1. This essay grew out of discussions with Dr. Paul Oppenheim; it was published in coauthorship with him and is here reprinted with his permission. Our individual contributions cannot be separated in detail; the present author is responsible, however, for the substance of Part IV and for the final formulation of the entire text.

Some of the ideas set forth in Part II originated with our common friend, Dr. Kurt Grelling, who suggested them to us in a discussion carried on by correspondence. Grelling and his wife subsequently became victims of the Nazi terror during the Second World War; by including in this essay at least some of Grelling's contributions, which are explicitly identified, we hope to realize his wish that his ideas on this subject might not entirely fall into oblivion.

Paul Oppenheim and I are much indebted to Professors Rudolf Carnap, Herbert Feigl, Nelson Goodman, and W. V. Quine for stimulating discussions and constructive criticism.

2. These two expressions, derived from the Latin *explanare*, were adopted in preference to the perhaps more customary terms "explicandum" and "explicans" in order to reserve the latter for use in the context of explication of meaning, or analysis. On explication in this sense, cf. Carnap (1945a), p. 513.

3. (Added in 1964.) Requirement (R4) characterizes what might be called a correct or *true explanation*. In an analysis of the logical structure of explanatory arguments, therefore, the requirement may be disregarded. This is, in fact, what is done in section 7, where the concept of *potential explanation* is introduced. On these and related distinctions, see also section 2.1 of the essay "Aspects of Scientific Explanation."

4. (Added in 1964.) This claim is examined in much fuller detail, and reasserted with certain qualifications, in sections 2.4 and 3.5 of the essay "Aspects of Scientific Explanation."

5. (1924), p. 533.

6. (Added in 1964.) Or rather, causal explanation is one variety of the deductive type of explanation here under discussion; see section 2.2 of "Aspects of Scientific Explanation."

7. The account given above of the general characteristics of explanation and prediction in science is by no means novel; it merely summarizes and states explicitly some fundamental points which have been recognized by many scientists and methodologists.

Thus, for example, Mill says: "An individual fact is said to be explained, by pointing out its cause, that is, by stating the law or laws of causation, of which its production is an instance", and "a law or uniformity in nature is said to be explained, when another law or laws are pointed out, of which that law itself is but a case, and from which it could be deduced." (1858, Book III, Chapter XII, section 1). Similarly, Jevons, whose general characterization of explanation was critically discussed above, stresses that "the most important process of explanation consists in showing that an observed fact is one case of a general law or tendency." (1924, p. 533). Ducasse states the same point as follows: "Explanation essentially consists in the offering of a hypothesis of fact, standing to the fact to be explained as case of antecedent to case of consequent of some already known law of connection." (1925, pp. 150–51). A lucid analysis of the fundamental structure of explanation and prediction was given by Popper in (1935), section 12, and, in an improved version, in his work (1945), especially in Chapter 25 and in note 7 for that chapter.—For a recent characterization of explanation as subsumption under general theories, cf., for example, Hull's concise discussion in (1943a), chapter 1. A clear elementary examination of certain aspects of explanation is given in Hospers (1946), and a concise survey of many of the essentials of scientific explanation which are considered in the first two parts of the present study may be found in Feigl (1945), pp. 284ff.

8. On the subject of explanation in the social sciences, especially in history, cf. also the following publications, which may serve to supplement and amplify the brief discussion to be presented here: Hempel (1942); Popper (1945); White (1943); and the articles "Cause" and "Understanding" in Beard and Hook (1946).

9. The illustration is taken from Bonfante (1946), section 3.

10. While in each of the last two illustrations, certain regularities are unquestionably relied upon in the explanatory argument, it is not possible to argue convincingly that the intended laws, which at present cannot all be stated explicitly, are of a causal rather than a statistical character. It is quite possible that most or all of the regularities which will be discovered as sociology develops will be of a statistical type. Cf., on this point, the suggestive observations in Zilsel (1941), section 8, and (1941a). This issue does not affect, however, the main point we wish to make here, namely that in the social no less than in the physical sciences, subsumption under general regularities is indispensable for the explanation and the theoretical understanding of any phenomenon.

11. Cf., for example, F. H. Knight's presentation of this argument in (1924), pp. 251–52.

12. For a detailed logical analysis of the concept of motivation in psychological theory, see Koch (1941). A stimulating discussion of teleological behavior from the standpoint of contemporary physics and biology is contained in the article (1943) by Rosenblueth, Wiener, and Bigelow. The logic of explanation by motivating reasons is examined more fully in Section 10 of my (1966) essay.

13. An analysis of teleological statements in biology along these lines may be found in Woodger (1929), especially pp. 432ff.; essentially the same interpretation is advocated by Kaufmann in (1944), Chapter 8.

14. For a more detailed discussion of this view on the basis of the general principles outlined above, cf. Zilsel (1941), sections 7 and 8, and Hempel (1942), section 6.

15. For a lucid brief exposition of this idea, see Feigl (1945), pp. 284–88.

16. Concerning the concept of novelty in its logical and psychological meanings, see also Stace (1939).

17. This observation connects the present discussion with a basic issue in Gestalt theory. Thus, for example, the insistence that "a whole is more than the sum of its parts" may be construed as referring to characteristics of wholes whose prediction requires knowledge of certain structural relations among the parts. For a further examination of this point, see Grelling and Oppenheim (1937–38) and (1939).

18. Logical analyses of emergence which make reference to the theories available have been propounded by Grelling and recently by Henle (1942). In effect, Henle's definition characterizes a phenomenon as emergent if it cannot be predicted, by means of the theories accepted at the time, on the basis of the data available before its occurrence. In this interpretation of emergence, no reference is made to characteristics of parts or constituents. Henle's concept of predictability differs from the one implicit in our discussion (and made explicit in Part III of this article) in that it implies derivability from the "simplest" hypothesis which can be formed on the basis of the data and theories available at the time. A number of suggestive observations on the idea of emergence and on Henle's analysis of it are presented in Bergmann's article (1944). The idea that the concept of emergence, at least in some of its applications, is meant to refer to unpredictability by means of "simple" laws was advanced also by Grelling in the correspondence mentioned in note (1). Reliance on the notion of simplicity of hypotheses, however, involves considerable difficulties; in fact, no satisfactory definition of that concept is available at present.

19. C. D. Broad, who in Chapter 2 of his book (1925) gives a clear account and critical discussion of the essentials of emergentism, emphasizes the importance of "laws" of composition in predicting the characteristics of a whole on the basis of those of its parts (op. cit., pp. 61ff.); but he subscribes to the view characterized above and illustrates it specifically by the assertion that "if we want to know the chemical (and many of the physical) properties of a chemical compound, such as silver-chloride, it is absolutely necessary to study samples of *that particular compound*. . . . The essential point is that it would also be useless to study chemical compounds in general and to compare their properties with those of their elements in the hope of discovering a *general* law of composition by which the properties of *any* chemical compound could be foretold when the properties of its separate elements were known." (p. 64) That an achievement of precisely this sort has been possible on the basis of the periodic system of the elements is noted above.

20. The following passage from Tolman (1932) may serve to support this interpretation. ". . . 'behavior-acts,' though no doubt in complete one-to-one correspondence with the underlying molecular facts of physics and physiology, have, as 'molar' wholes, certain emergent properties of their own. . . . Further, these molar properties of behavior-acts cannot in the present state of our knowledge, that is, prior to the working-out of many empirical correlations between behavior and its physiological correlates, be known even inferentially from a mere knowledge of the underlying, molecular, facts of physics and physiology" (op. cit., pp. 7–8). In a similar manner, Hull uses the distinction between molar and molecular theories and points out that theories of the latter type are not at present available in psychology. Cf. (1943a), pp. 19ff.; (1943), p. 275.

21. This attitude of the scientist is voiced, for example, by Hull in (1943a), pp. 24–28.

22. The requirement of truth for laws has the consequence that a given empirical statement S can never be definitely known to be a law; for the sentence affirming the truth of S is tantamount to S and is therefore capable only of acquiring a more or less high probability, or degree of confirmation, relative to the experimental evidence available at any given time. On this point, cf. Carnap (1946). For an excellent nontechnical exposition of the semantical concept of truth, which is here invoked, the reader is referred to Tarski (1944).

23. (1947), p. 125.

24. This procedure was suggested by Goodman's approach in (1947). Reichenbach, in a detailed examination of the concept of law, similarly construes his concept of nomological statement as including both analytic and synthetic sentences: cf. (1947). Chapter VIII.

25. The difficulty illustrated by this example was stated concisely by Langford (1941), who referred to it as the problem of distinguishing between universals of fact and causal universals. For further discussion and illustration of this point, see also Chisholm (1946), especially pp. 301f. A systematic analysis of the problem was given by Goodman in (1947), especially part III. While not concerned with the specific point under discussion, the detailed examination of counterfactual conditionals and their relation to laws of nature, in Chapter VIII of Lewis (1946), contains important observations on several of the issues raised in the present section.

26. The view that laws should be construed as not being limited to a finite domain has been expressed, among others, by Popper (1935), section 13, and by Reichenbach (1947), p. 369.

27. (1947), p. 361. Our terminology as well as the definitions to be proposed later for the two types of law do not coincide with Reichenbach's, however.

28. In physics, the idea that a law should not refer to any particular object has found its expression in the maxim that the general laws of physics should contain no reference to specific space-time points, and that spatio-temporal coordinates should occur in them only in the form of differences or differentials.

29. The point illustrated by the sentences S_3 and S_4 above was made by Goodman, who has also emphasized the need to impose certain restrictions upon the predicates whose occurrence is to be permissible in lawlike sentences. These predicates are essentially the same as those which Goodman calls projectible. Goodman has suggested that the problems of establishing precise criteria for projectibility, of interpreting counterfactual conditionals, and of defining the concept of law are so intimately related as to be virtually aspects of a single problem. Cf. his articles (1946) and (1947). One suggestion for an analysis of projectibility has been made by Carnap in (1947). Goodman's note (1947a) contains critical observations on Carnap's proposals.

30. That laws, in addition to being of universal form, must contain only purely universal predicates was argued by Popper (1935, sections 14, 15). Our alternative expression 'purely qualitative predicate' was chosen in analogy to Carnap's term 'purely qualitative property' cf. (1947). The above characterization of purely universal predicates seems preferable to a simpler and perhaps more customary one, to the effect that a statement of the meaning of the predicate must require no reference to particular objects. That formulation might be too restrictive since it could be argued that stating the meaning of such purely qualitative terms as 'blue' or 'hot' requires illustrative reference to some particular object which has the quality in question. The essential point is that no one specific object has to be chosen; any one in the logically unlimited set of blue or of hot objects will do. In explicating the meaning of 'taller than the Eiffel Tower', 'being an apple in basket b at time t', 'medieval', etc., however, reference has to be made to one specific object or to some one in a limited set of objects.

31. As defined above, fundamental laws include universal conditional statements with vacuous antecedents, such as "All mermaids are brunettes." This point does not appear to lead to undesirable consequences in the definition of explanation to be proposed later. For an illuminating analysis of universal conditionals with vacuous antecedents, see Chapter VIII in Reichenbach (1947).

32. (Added in 1964.) However, Nagel has shown that our definition of the concept of fundamental law is too restrictive; cf. the postscript to the present essay.

33. This is not a matter of free choice: The precise rational reconstruction of explanation as applied to general regularities presents peculiar problems for which we can offer no solution at present. The core of the difficulty can be indicated briefly by reference to an example: Kepler's laws, K, may be conjoined with Boyle's law, B, to a stronger law $K \cdot B$; but derivation of K from the latter would not be considered as an explanation of the regularites stated in Kepler's laws; rather, it would be viewed as representing, in effect, a pointless "explanation" of Kepler's laws by themselves. The derivation of Kepler's laws from Newton's laws

of motion and of gravitation, on the other hand, would be recognized as a genuine explanation in terms of more comprehensive regularities, or so-called higher-level laws. The problem therefore arises of setting up clear-cut criteria for the distinction of levels of explanation or for a comparison of generalized sentences as to their comprehensiveness. The establishment of adequate criteria for this purpose is as yet an open problem.

34. It is necessary to stipulate, in (7.6) (2), that T be a theory rather than merely that T be true; for as was shown in section 5, the generalized sentences occurring in an explanans have to constitute a theory, and not every esentially generalized sentence which is true is actually a theory, that is, a consequence of a set of purely generalized true sentences.

35. In the formulation of the above theorem and subsequently, statement connective symbols are used not only as signs in L, but also autonymously in speaking about compound expressions of L. Thus, when 'S' and 'T' are names or name variables for sentences in L, their conjunction and disjunction will be designated by '$S \cdot T$' and 'SvT', respectively; the conditional which has S as antecedent and T as consequent will be designated by '$S \supset T$', and the negation of S by '$\sim S$'. (Incidentally, this convention has already been used, tacitly, at one place in note 33).

36. The characteristic here referred to as partial self-explanation has to be distinguished from what is sometimes called the circularity of scientific explanation. The latter phrase has been used to cover two entirely different ideas. (a) One of these is the contention that the explanatory principles adduced in accounting for a specific phenomenon are inferred from that phenomenon, so that the entire explanatory process is circular. This belief is false, since general laws cannot be inferred from singular sentences. (b) It has also been argued that in a sound explanation the content of the explanandum is contained in that of the explanans. That is correct since the explanandum is a logical consequence of the explanans; but this peculiarity does not make scientific explanation trivially circular since the general laws occurring in the explanans go far beyond the content of the specific explanandum. For a fuller discussion of the circularity objection, see Feigl (1945), pp. 286ff, where this issue is dealt with very clearly.

37. It is important to distinguish clearly between the following two cases: (a) If T is true then C cannot be true without E being true; and (b) If T is true, C cannot be verified without E being verified. Condition (a) must be satisfied by any potential explanation; the much more restrictive condition (b) must not be satisfied if (T,C) is to be a potential explanans for E.

38. As this stipulation shows, the term "datum" is here understood as covering actual as well as potential data. The convention that any singular sentence expresses a potential datum is plausible especially if the primitive predicates of L refer to attributes whose presence or absence in specific instances can be ascertained by direct observation. In this case each singular sentence in L may be considered as expressing a potential datum in the sense of describing a logically possible state of affairs whose existence might be ascertained by direct observation. The assumption that the primitives of L express directly observable attributes is, however, not essential for the definition and the formal theory of systmatic power set forth in sections 8 and 9.

39. Cf. especially (1945), (1945a), (1947).

40. See Helmer and Oppenheim (1945); Hempel and Oppenheim (1945). Certain general aspects of the relationship between the confirmation of a theory and its predictive or systematic success are examined in Hempel (1945), Part II, sections 7 and 8. The definition of s developed in the present essay establishes a quantitative counterpart of what, in that paper, is characterized, in non-numerical terms, as the prediction criterion of confirmation.

41. For a definition and discussion of this concept, see, for example, Church (1942), p. 172.

42. A more detailed discussion of the concept of range may be found in Carnap (1945), section 2, and in Carnap (1942), sections 18 and 19, where the relation of range and content is examined at length.

43. In Carnap's theory of logical probability, $p(T,S)$ is defined, for certain cases, as the limit which the function $r(T \cdot S)/r(S)$ assumes under specified conditions, cf. Carnap (1945), p. 75; but we shall refrain here from considering this generalization of that type of definition which is represented by (9.3).

44. (1945), pp. 80–81.

45. The alternative approach suggested by Helmer and the present authors involves use of a range measure function r_1 which depends in a specified manner on the empirical information I available; hence, the range measure of any sentence U is determined only if a sentence I, expressing the available empirical information, is given. In terms of this range measure function, the concept of degree of confirmation, dc, can be defined by means of a formula similar to (9.3). The value of $dc(T,S)$ is not defined, however, in certain cases where S is generalized, as has been pointed out by McKinsey (1946); also, the concept dc does not satisfy all the theorems of elementary probability theory (cf. the discussion of this point in the first two articles mentioned in note 40); therefore, the degree of confirmation of a theory relative to given evidence is not a probability in the strict sense of the word. On the other hand, the definition of dc here referred to has certain metholodogically desirable features, and it might therefore be of interest to construct a related concept of systematic power by means of the range measure function r_1. In the present paper, however, this question will not be pursued.

46. Cf. especially Goodman (1946), (1947), (1947a) and Carnap (1947).

References

Beard, Charles A., and Sidney Hook, "Problems of Terminology in Historical Writing." Chapter IV of *Theory and Practice in Historical Study: A Report of the Committee on Historiography*. New York, Social Science Research Council, 1946.

Bergmann, Gustav, "Holism, Historicism, and Emergence," *Philosophy of Science*, 11 (1944), 209–21.

Bonfante, G., "Semantics, Language," in P. L. Harriman, ed., *The Encyclopedia of Psychology*. New York, 1946.

Broad, C. D., *The Mind and Its Place in Nature*. New York, 1925.

Carnap, Rudolf, *Introduction to Semantics*. Cambridge, Mass., 1942.

———, "On Inductive Logic," *Philosophy of Science*, 12 (1945), 72–97.

———, "The Two Concepts of Probability," *Philosophy and Phenomenological Research*, 5 (1945), 513–32.

———, "Remarks on Induction and Truth," *Philosophy and Phenomenological Research*, 6 (1946), 590–602.

———, "On the Application of Inductive Logic," *Philosophy and Phenomenological Research*, 8 (1947), 133–47.

Chisholm, Roderick M., "The Contrary-to-Fact Conditional," *Mind*, 55 (1946), 289–307.

Church, Alonzo, "Logic, formal," in Dagobert D. Runes, ed. *The Dictionary of Philosophy*. New York, 1942.

Ducasse, C. J., "Explanation, Mechanism, and Teleology," *The Journal of Philosophy*, 22 (1925), 150–55.

Feigl, Herbert, "Operationism and Scientific Method," *Psychological Review*, 52 (1945), 250–59, 284–88.

Goodman, Nelson, "A Query on Confirmation," *The Journal of Philosophy*, 43 (1946), 383–85.

———, "The Problem of Counterfactual Conditionals," *The Journal of Philosophy*, 44 (1947), 113–28.

———, "On Infirmities of Confirmation Theory," *Philosophy and Phenomenological Research*, 8 (1947), 149–51.

Grelling, Kurt, and Paul Oppenheim, "Der Gestaltbegriff im Lichte der neuen Logik," *Erkenntnis,* 7 (1937–38), 211–25, 357–59.

———, "Logical Analysis of Gestalt as 'Functional Whole'." Preprinted for distribution at Fifth International Congress for the Unity of Science, Cambridge, Mass., 1939.

Helmer, Olaf, and Paul Oppenheim, "A Syntactical Definition of Probability and of Degree of Confirmation," *The Journal of Symbolic Logic,* 10 (1945), 25–60.

Hempel, Carl G., "The Function of General Laws in History." *The Journal of Philosophy,* 39 (1942), 35–48.

——— "Studies in the Logic of Confirmation," *Mind,* 54 (1945); Part I: pp. 1–26, Part II: pp. 97–121.

——— "Aspects of Scientific Explanation," in *Aspects of Scientific Explanation and Other Essays in the Philosophy of Science.* New York: The Free Press, 1965, pp. 331–496.

Hempel, Carl G., and Paul Oppenheim, "A Definition of Degree of Confirmation," *Philosophy of Science,* 12 (1945), 98–115.

Henle, Paul, "The Status of Emergence," *The Journal of Philosophy,* 39 (1942), 486–93.

Hospers, John, "On Explanation," *The Journal of Philosophy,* 43 (1946), 337–56.

Hull, Clark L., "The Problem of Intervening Variables in Molar Behavior Theory," *Psychological Review,* 50 (1943), 273–91.

———, *Principles of Behavior.* New York, 1943.

Jevons, W. Stanley, *The Principles of Science.* London, 1924 (1st ed. 1874).

Kaufmann, Felix, *Methodology of the Social Sciences.* New York, 1944.

Knight, Frank H., "The Limitations of Scientific Method in Economics," in R. Tugwell, ed., *The Trend of Economics.* New York, 1924.

Koch, Sigmund, "The Logical Character of the Motivation Concept," *Psychological Review,* 48 (1941); Part I: pp. 15–38, Part II: pp. 127–54.

Langford, C. H., Review in *The Journal of Symbolic Logic,* 6 (1941), pp. 67–68.

Lewis, C. I., *An Analysis of Knowledge and Valuation.* La Salle, Ill., 1946.

McKinsey, J. C. C., Review of Helmer and Oppenheim (1945). *Mathematical Review,* 7 (1946), p. 45.

Mill, John Stuart, *A System of Logic.* New York, 1858.

Morgan, C. Lloyd, *Emergent Evolution.* New York, 1923.

———, *The Emergence of Novelty.* New York, 1933.

Popper, Karl, *Logik der Forschung.* Wien, 1935.

———, *The Open Society and Its Enemies.* London, 1945.

Reichenbach, Hans, *Elements of Symbolic Logic.* New York, 1947.

Rosenblueth, A., N. Wiener, and J. Bigelow, "Behavior, Purpose, and Teleology," *Philosophy of Science,* 10 (1943), 18–24.

Stace, W. T., "Novelty, Indeterminism and Emergence," *Philosophical Review,* 48 (1939), 296–310.

Tarski, Alfred, "The Semantical Conception of Truth, and the Foundations of Semantics," *Philosophy and Phenomenological Research,* 4 (1944), 341–76.

Tolman, Edward Chase, *Purposive Behavior in Animals and Men.* New York, 1932.

White, Morton G., "Historical Explanation." *Mind,* 52 (1943), 212–29.

Woodger, J. H., *Biological Principles.* New York, 1929.

Zilsel, Edgar, *Problms of Empiricism.* Chicago, 1941.

———, "Physics and the Problem of Historico-Sociological Laws," *Philosophy of Science,* 8 (1941), 567–79.

Postscript (1964) to Studies in the Logic of Explanation

Carl G. Hempel

The preceding essay has been widely discussed in the philosophical literature. Most of the discussion has been concerned with the general conception, set forth in Part I, of explanation by deductive subsumption under laws or theoretical principles. Indeed, some commentators seem to attribute to me the view that all adequate scientific explanations must be of this type, despite the fact that in the final paragraphs of sections 3 and 7 of the essay, as well as in section 5.3 of the earlier article "The Function of General Laws in History,"[1] another type of explanation is acknowledged, which invokes probabilistic-statistical laws. The logic of such explanation is not, however, further explored in either of those two articles; an attempt to fill this gap is made in section 3 of the essay "Aspects of Scientific Explanation."[2] That essay also incorporates my responses to some of the stimulating comments and criticisms that have been directed at the two earlier studies.

In this Postscript, I will limit myself to surveying certain shortcomings of the ideas developed in Part III of the preceding essay.

(1) As E. Nagel has rightly pointed out,[3] the definition (7.3b) of the concept of derivative law is too restrictive; for, contrary to the intention indicated in section 6, it bars such laws as Galileo's and Kepler's from the status of derivative laws. This is so because those generalizations cannot be derived from the fundamental Newtonian laws of mechanics and of gravitation alone—which, in effect, would have to be done solely by substituting constant terms for variables occurring in the latter. Actually, the derivation requires additional premises which do not have the character of fundamental laws; in the case of Galileo's law, for example, these include statements specifying the mass and the radius of the earth. (In fact, even with the help of such additional premises, Galileo's and Kepler's laws cannot strictly be derived from the Newtonian principles; they are only approximations of statements that are so derivable. However, this point, which is discussed further in section 2 of "Aspects of Scientific Explanation," clearly does not diminish the force of Nagel's argument.)

Nagel notes further that if the definition (7.3b) were modified so as to countenance the use of additional non-lawlike premises, then certain unfit candidates would be qualified as derivative laws. Indeed, this would be true, for example, of the sentence 'Every apple now in this basket is red', which is deducible from the (putative) law 'All Winesap apples are red' in conjunction with the premise 'Every apple now in this basket is a Winesap.' Nagel illustrates his point by the sentence 'All screws in Smith's car are rusty', which is deducible from the law 'All iron exposed to oxygen rusts', in conjunction with suitable particular premises.

What bars generalizations like the two just mentioned from the status of potential laws appears to be their limited scope: each seems to pertain to only a finite number of objects. This observation suggests that the requirement of nonlimitation of scope, which in section 6 was imposed on fundamental lawlike sentences, should be extended to derivative lawlike sentences as well. And indeed, Nagel requires that lawlike sentences in general should be "unrestricted universals," that is, that their "scope of predication" must not fall into "a fixed spatial region or a particular period of time."[4] But on this formulation of the intended requirement, it may happen that a given sentence is disqualified whereas another, logically equivalent one is not. For example, the two restricted universals just considered are logically equivalent to the following generalizations, whose scopes of predication clearly do satisfy Nagel's condition: 'Anything that is not red is not an apple in this basket' and 'Any object that is nonrusty is not a screw in Smith's car'.

This difficulty is avoided if the scope requirement is given the following form: Except for purely logical truths (which are equivalent to '$Pa \vee \sim Pa$'), lawlike statements must not have a finite scope in the sense of being logically equivalent to some finite conjunction of singular sentences about particular cases (as in 'apple a is red and apple b is red and apple c is red'); or, more precisely and briefly: they must be essentially universal. Evidently, if a sentence satisfies this condition then so does any logical equivalent of it.

This condition, which the definitions (7.3a) and (7.3b) do in fact impose upon fundamental and derivative lawlike sentences, is discussed more fully in section 2.1 of "Aspects of Scientific Explanation." But while clearly a necessary condition for lawlike sentences, it is too weak fully to avoid the difficulty pointed out by Nagel. In fact, it does not rule out the two undesirable generalizations just considered: neither of these can be equivalently transformed into a finite conjunction of singular sentences about particular apples or screws, for the sentences do not even indicate how many apples there are in the basket or how many screws in Smith's car; and even less do they provide a list of names for the individual objects referred to, as would be required for the transformation. Hence it remains an important desideratum to find a satisfactory version of the scope condition which requires more of a lawlike sentence than that it must be essentially universal.

(2) I now turn to a shortcoming of the definition (7.8) of a potential explanans. That definition, as I realized a number of years ago, is much too inclusive; for, in a sense presently to be illustrated, it countenances the explanation of any particular fact by itself and makes it possible to generate a potentially explanatory theory for any given particular fact from any essentially generalized sentence. Consider, for example, the argument

$$\text{(2a)} \qquad \begin{array}{c} (x)Px \\ \underline{Qa} \\ Qa \end{array} \quad \text{or briefly} \quad \begin{array}{c} T \\ \underline{C} \\ C \end{array}$$

It has the form of a complete self-explanation and is therefore ruled out by condition (3) in definition (7.8). But its explanans can be equivalently restated in a form which is acceptable under (7.8), and which yields the following argument:

(2b) $$\frac{(x)(Px \cdot Qa)}{\underline{Qa \vee \sim Qa}}{Qa}$$ or briefly $$\frac{T'}{\underline{C'}}{C}$$

This argument clearly satisfies conditions (1) and (2) in (7.8). But it also satisfies condition (3); for T' is compatible with the class containing the basic sentence 'Pb' as its only element; and that class has C' but not C as a logical consequence.

This flaw can be eliminated by limiting T in definition (7.8) to purely generalized sentences. However, this is a highly undesirable restriction, for the definition was also intended to cover explanation by means of derivative laws and theories.

(3) But even if one were willing to pay this price, the modified version of (7.8) would still have quite unacceptable consequences. This has been brought to light in an incisive critical study by Eberle, Kaplan, and Montague,[5] which shows that virtually any fundamental theory yields an explanation in the sense of (7.8) for virtually any particular fact. The authors establish this by proving five theorems, each of which exhibits such explainability relations for some large class of cases in which the theory would normally be regarded as irrelevant to the fact to be explained.

The first of those theorems, for example, is this: Let T be a fundamental law and E a true singular sentence, neither of which is logically provable in the language L, and furthermore, let the two sentences have no predicates in common—so that, intuitively speaking, T deals with a subject matter totally different from that of E. Then, granting only the availability of an adequate supply of further individual constants and predicates in L, there is a fundamental law T' which is logically derivable from T and by which E is explainable in the sense of definition (7.9). For example, let T be '$(x)Fx$' and E be 'Ha'; then consider the sentence

$$T' : (x)(y)[Fx \vee (Gy \supset Hy)]$$

It is of purely universal form and is derivable from T and thus is true since by hypothesis T is a law and therefore true. Hence, T' is a fundamental law. Next, consider the sentence

$$C : (Fb \vee \sim Ga) \supset Ha$$

This sentence is singular and is a consequence of E and thus is true, since by hypothesis, E is true. And as can now readily be verified, (T',C) forms a potential explanans (and indeed a true one) for E in the sense of (7.8).

I am happy to be able to say in conclusion that it is possible to modify the definitions (7.8) and (7.9) so as to forestall these disabling consequences. One method has been pointed out by one of the authors of the critical study just discussed, D. Kaplan.[6] An alternative modification has been devised by J. Kim.[7]

The crucial part of Kim's revision is a requirement to be added to those specified in (7.8), to the following effect: Let C be put into a complete conjunctive normal form in those atomic sentences which occur essentially in C; then none of the conjuncts of that normal form must be logically derivable from it. In our illustration of the first of the five critical theorems, this requirement is violated; for 'Ha' logically implies, in fact, every one of the conjuncts of the complete conjunctive normal form of '$(Fb \vee \sim Ga) \supset Ha$', namely, '$Fb \vee Ga \vee Ha$', '$\sim Fb \vee Ga \vee Ha$', and

'~*Fb* v ~ *Ga* v *Ha*'. Kim shows generally that his additional requirement blocks the proofs offered by Eberle, Kaplan, and Montague for the five theorems that "trivialize" the defintions (7.8) and (7.9). However, it would be desirable to ascertain more clearly to what extent the additional requirement is justifiable, not on the *ad hoc* ground that it blocks those proofs, but in terms of the rationale of scientific explanation.

Kaplan approaches the problem by formulating three very plausible requirements c f adequacy for any analysis of the deductive type of explanation here to be explicated. He then shows that the analysis proposed in Part III does not satisfy those requirements jointly, and that the difficulties exhibited in the five trivializing theorems are linked to this shortcoming. Finally, he revises the definitions offered in Part III so that they meet the requirements of adequacy and avoid the difficulties we have been discussing. For the details of this illuminating contribution, the reader will have to consult Kaplan's article.

Notes

1. C. Hempel, "The Function of General Laws in History," in *Aspects of Scientific Explanation and Other Essays in the Philosophy of Science*. New York: The Free Press, 1965, pp. 35–48.

2. Hempel, "Aspects of Scientific Explanation," op. cit., pp. 331–496.

3. E. Nagel, *The Structure of Science*. New York, 1961, p. 58.

4. Op. cit., p. 59.

5. R. Eberle, D. Kaplan, and R. Montague, "Hempel and Oppenheim on Explanation," *Philosophy of Science* 28 (1961), pp. 418–28.

6. D. Kaplan, "Explanation Revisited," *Philosophy of Science* 28 (1961), pp. 429–36.

7. J. Kim, "Discussion: On the Logical Conditions of Deductive Explanation," *Philosophy of Science* 30 (1963), pp. 286–91.

3

Explanations, Predictions, and Laws

Michael Scriven

3. Preliminary Issues

3.1. Explanations: Introduction

I am going to take a series of suggested analytical claims about the logic of explanation and gradually develop a general idea of what is lacking in them, or too restrictive about them. I shall at each stage try to formulate criteria which will survive the difficulties while retaining the virtues of the current candidate. Eventually I shall try to draw the surviving criteria together into an outline of a new account of both explanation and understanding; but this will not be possible until I have encompassed the whole field of topics envisioned above. The questions with which I begin will seem quite unimportant; but they are in fact more significant than they appear because of the cumulative error in the standard answers to them.

3.2. Explanations as Answers to "Why" Questions

"To explain the phenomena in the world of our experience, to answer the question 'Why?' rather than only the question 'What?' . . ." With these words, Hempel and Oppenheim begin their monograph on scientific explanation.[1] Braithwaite says, "Any proper answer to a 'Why?' question may be said to be an explanation of a sort. So the different kinds of explanation can best be appreciated by considering the different sorts of answers that are appropriate to the same or to different 'Why?' questions."[2]

This happens to be a *non sequitur*, but it is the conclusion that I particularly wish to consider. The "answer-to-a-why-question" criterion could have been proposed in the absence of a serious attempt to think of counterexamples. "How can a neutrino be detected, when it has zero mass and zero charge?" is a perfectly good request for a perfectly standard scientific explanation. "What is it about cepheid variables that makes them so useful for the determination of interstellar distances?" Likewise, the same can be said of suitable Which, Whither, and When questions.

Not all, perhaps not even most, of the answers to such questions involve explanations, whereas it is perhaps true that most answers to Why questions are explanations. (But not all, as for example an answer that rebuts a presupposition of the question "Why do you persist in lying?" "I have *never* lied about this affair.") But explanations are also given when no questions are asked at all, as in the course of a lecture, or in correcting or supporting an assertion. The identifying feature of an explanation cannot, therefore, be the grammatical form of the question which (sometimes) produces it. Indeed, it is fairly clear that one does not teach a foreigner or a child the word "explanation" simply by reference to Why questions; so the authors quoted presumably had some prior (or at least alternative) notion of explanation in mind which enabled them to identify answers to Why questions as explanations. Should we not look for the meaning of that notion?

It is sometimes replied that our common notion of explanation is excessively vague, and it is therefore quite unrewarding to seek its exact meaning; far better to concentrate on some substantial concept which clearly does occur. This is a very good reply and represents a sensible approach, if only it can be shown to be true. This requires showing (a) that the ordinary notion *is* excessively vague, and (b) that the "substantial" alternative occurs often enough to justify any general conclusions about explanations which are inferred from studying it. I shall be arguing that neither of these seemingly innocuous premises can be established, and in consequence the analysis suggested by the reconstructionish authors is fundamentally unsatisfactory.

Explanation is undoubtedly a notion whose analysis must be sought in the practical foundation of language; but it is too much to hope one can identify explanations by such a simple linguistic device as the one suggested. Nor will it do to suppose that explanations are such that they are answers to *potential* Why questions; for then they are also potentially answers to What-about questions, How-possibly questions, etc. Thus, to take an example quoted by Hempel and Oppenheim, the question "Why does a stick half-immersed in water appear bent?" can readily be rephrased as "What makes a stick (in such circumstances) seem to be bent?" Indeed, such a question as "How can the sun possibly continue to produce so much energy with a negligible loss of mass?" is only with some difficulty rephrased as a Why question.[3] In sum, the grammatical indicators of explanations are complicated and none of them are necessary; some more illuminating and reliable criteria must be sought.

3.3. *Explanations as "More Than" Descriptions*

Another common remark in the literature is that explanations are more than descriptions. This is put by Hempel and Oppenheim in the following words: ". . . especially, scientific research in its various branches strives to go beyond a mere description of its subject matter by providing an explanation of the phenomena it investigates."[4] But if one goes on to examine their own examples of explanations one finds what seem to be simply complex descriptions.[5] Thus they offer an explanation of the fact that when "a mercury thermometer is rapidly immersed in hot water, there occurs a temporary drop of the mercury column, which is then followed by a swift rise." And the explanation consists of the following account: "The increase in temperature affects at first only the glass tube of the thermometer; it expands and thus provides a larger space for the mercury inside, whose surface therefore drops. As soon as by heat conduction the rise in temperature reaches the mercury, however, the latter expands, and as its coefficient of expansion is considerably larger than that of glass, a rise of the mercury level results."[6]

This is surely intended to be a narrative description of exactly what happens. The one feature which might suggest a difference from a "mere description" is the occurrence of such words as "thus," "however," and "results." These are *reminiscent* of an argument or demonstration, and I think partially explain the analysis proposed by Hempel and Oppenheim, and others. But they are not part of an argument or demonstration here, simply of an explanation; and they or their equivalents occur in some of the simplest descriptions. "The curtains knocked over the vase" is a description which includes a causal claim and it could equally well be put, style aside, as "The curtains brushed against the vase, *thus* knocking it over" (or ". . . resulting in it being knocked over"). The fact that it is an explanatory account is therefore not in any way a ground for saying it is not a descriptive account (cf. "historical narrative"). Indeed, if it was not descriptive of what happens, it could hardly be explanatory. The question we have to answer is how and when certain descriptions count as explanations. Explaining how fusion processes enable the sun to maintain its heat output consists exactly in describing these processes and their products. Explaining therefore sometimes consists simply in giving the *right* description. What counts as the right description? Tentatively we can consider the vague hypothesis that the right description is the one which fills in a particular gap in the understanding of the person or people to whom the explanation is directed. That there is a gap in understanding, or a misunderstanding, seems plausible since whatever an explanation *actually* does, in order to be called an explanation at all it must be *capable* of making clear something not previously clear, that is, of increasing or producing understanding of something. The difference between explaining and "merely" informing, like the difference between explaining and describing, does not, I shall argue, consist in explaining being something "more than" or even something intrinsically different from informing or describing, but in its being the appropriate piece of informing or describing, the appropriateness being a matter of its relation to a *particular context*. Thus, what would in one context be "a mere description" can in another be "a full explanation." The distinguishing features will be found, not in the verbal form of the question or answer, but in the known or inferred state of understanding and the proposed explanation's relation to it. To these, of course, the form of the question and answer are often important clues, though not the only clues. But this is only a rough indication of the *direction* of the solution to be proposed in this paper, and it may be that the notion of understanding will present us with substantial difficulties, quite apart from the problem of identifying the criteria for "closing the gap" in understanding (or rectifying the misunderstanding). However, let me remind the reader that understanding is *not* a subjectively appraised state any more than knowing is; both are objectively testable and are, in fact, tested in examinations. We may first benefit from examining the relation between explanation and another important scientific activity.

3.4. Explanations as "Essentially Similar" to Predictions

The next suggestion to be considered is a much more penetrating one, and although it cannot be regarded as satisfactory, the reasons for dissatisfaction are more involved. Quoting from Hempel and Oppenheim once more: ". . . the same formal analysis . . . applies to scientific prediction as well as to explanation. The difference between the two is of a pragmatic character . . . It may be said, therefore, that an explanation is not fully adequate unless . . . if taken account of in time, [it] could have served as a basis for predicting the phenomenon under consideration."[7]

(3.41) The full treatment of this view will require some points that will only be made later in the paper; but we can begin with several rather weighty objections. First, there certainly seem to be occasions when we can predict some phenomenon with the greatest success, but cannot provide any explanation of it. For example, we may discover that whenever cows lie down in the open fields by day, it always rains within a few hours. We are in an excellent position for prediction, but we could scarcely offer the earlier event as an explanation of the latter. It appears that explanation requires something "more than" prediction; and my suggestion would be that, whereas an understanding of a phenomenon often enables us to forecast it, the ability to forecast it does not constitute an understanding of a phenomenon.

(3.42) Indeed, the forecast is simply a description of an event (or condition, etc.) given prior to its occurrence and identified as referring to a future time; whereas an explanation will have to do more than merely describe those *features of the thing to be explained that identify it*. (In this sense, it is more than a (particular) description.[8]) At the very least some other features of it must be mentioned, and often some reference is made to previous or (other) concurrent events and/or laws. Since none of this is required of a prediction, it seems rather extraordinary to suppose that the *contents* of a prediction are logically identical to those of an explanation. And our first point showed that the *grounds* for the two are often quite different, in that one can be inferred from a mere correlation and the other not.

Such cases also demonstrate that explaining something is by no means the same as showing it was to be expected, since the latter task can be accomplished without any explanation being given.[9] For our purpose, however, the crucial point is that, however achieved, a prediction is what it is, simply because it is produced in advance of the event it predicts; it is *intrinsically* nothing but a bare description of that event. Whereas an explanation of the event *must* be more than the *identifying* description of it, else to request an explanation of X (where "X" is a description, not a name) is to give an explanation of X. Of course, there is usually a difference of tense, but we could agree to this as a pragmatic difference. However, it is the *least* and not the *only* difference between explanations and predictions.

(3.43) There also seem to be cases where explanation is not in terms of temporally ordered and causally related events, and we are consequently never able to make predictions. These cases are common enough outside the physical sciences, for example, in explaining the rules of succession in an Egyptian dynasty, or the symbolism of a tribal dance. Within science there are of course all the cases of explaining a theory or mechanism or proof; these are normally dismissed by supporters of the Hempel and Oppenheim position, on the grounds that they are clearly a different kind of explanation, the explanation of *meaning*, not at all related to *scientific* explanation. While there is no doubt about the difference in procedure between explaining a theory and explaining some phenomenon in terms of the theory, it is not enough to appeal to intuition for support of the claim that they are not "fundamentally," that is, for all logical purposes, the same except for subject matter, much as definition in mathematics might be said to be fundamentally the same as definition in the empirical sciences except for subject matter. In fact, it seems clear enough that one important element is held in common between the two "kinds" of explanation, viz. the provision of understanding. But is there not a great deal of difference between the kinds of understanding provided in the two cases?

Now, *not* understanding a theory may be due to not understanding what its assumptions are, to not understanding the meaning of some of its terms, or to not understanding how the derivations said to be possible from it are to be made. One might suppose this to be quite unlike not understanding why a stick half-immersed in water appears bent.

But instead of asking how we go about explaining a natural phenomenon, let us ask how we come to ask for an explanation, that is, what it is that we think *needs* explanation. It may seem that science is committed to the view that *everything* needs to be explained. Now it is clear that *everything* cannot be explained *every* time we give an explanation of some particular thing (or set of things) which is all we ever do in a given context. So we can rephrase our question as, What is it that needs explanation in a given context? It seems clear that it is those things which are not properly understood (by whomever the explanation is addressed to). Now, lack of understanding of a natural phenomenon may be due to the absence of certain information about the situation, to the presence of false beliefs about it, or to an inability to see the connections between what is understood and what is not understood. These are much the same kinds of difficulty as occur in not understanding a theory, although the information will be in one case about a verbal construction out of our knowledge and in the other about, for example, a mechanical construction out of our raw materials. However important the differences of subject may be, it is not obvious that the notion of understanding or explanation involved is in any important way different; and it is quite obvious that no predictions are possible on the basis of an explanation of the meaning of a theory (except, irrelevantly, those which the theory makes possible, if any—and it may be a theory whose advantages lie solely in its unifying powers). Certainly one should feel uneasy about any general claims of common logical structure for explanation and prediction which have to be defended by rejecting clear cases of explanation as "essentially different," without detailed examination. I shall argue that the differences are much less important than the similarities: in effect we are in both cases providing a series of comprehensible statements that have some of a wide range of logical relations to other statements. Lest this seem to be a proof of similarity by simply weakening the definition of 'similar,' I also try to show that the narrower definition is independently unsatisfactory.

(3.44) Again, we often talk of *explaining laws*: indeed half of Hempel and Oppenheim's examples are of this kind. Now, when we offer an explanation of Newton's Law of Cooling (that a body cools at a rate proportional to the difference between its temperature and that of its surroundings), we do so—according to Hempel and Oppenheim—by deriving this law from more general laws.[10] What predictions could be made which would have "the same formal structure" as this kind of explanation? The "pragmatic" difference between the two as they see it is essentially that explanation occurs after the phenomenon, and prediction before. But in the case of laws, which are presumably believed to hold at all times, what does it mean to talk of predicting the phenomenon? It is surely the case that the truth of Newton's law is *simultaneous* with that of the more general laws from which it is derivable. We cannot speak of being able to predict the inclusion of the class of A's in the class of C's if we already know that A's are B's and B's are C's.

It may seem that this argument can only be countered by saying that a law is a generalization about a number of events and that to "predict a law" is to predict the outcome of experiments done to determine the pattern of these events. It is true that this is different from predicting an eclipse, where the actual event to which the prediction refers is in the future, not merely the discovery of its nature. However, (i) it is certainly true that *some*—but not all—events governed by most laws lie entirely in the future, and its truth depends on these and these are predictable in the usual sense. (ii) Certainly, too, we want to say that inferences about the past, which *generate* predictions about what archaeologists or geologists will discover, have exactly the same logical structure as inferences about the future, a fact well brought out by the practice of calling them postdictions or retrodictions. So, if *explaining a law* consists in explaining the overall pattern of events, past, present, and future, *predicting a law*, as it seems we might interpret it, could be regarded as compounded out of such predictions and postdictions. This interpretation represents at the very least an *extension of meaning*, since we cannot in the usual sense call inferences about the activities of the earth's crust in Jurassic times (which will be covered by geological laws) predictions. Although we could quite properly apply this term to inferences about what will be found by geologists upon searching in certain areas, this is *not* what the law is about (for else we must say that the law asserts something different every time something new is discovered by the geologists, there being that much less for them still to discover). Indeed, we land in a well-known swamp if we make this move; for the same argument makes all historical statements into statements about contemporary evidence and all statements about distant places into statements about local evidence, etc. It is the argument which confuses the *reference of* a statement with the *evidence for* a statement. So explanations of laws only have a correlate among predictions if we *extend* the meaning of the notion of prediction to include postdiction.

Now this extension may not seem very significant until one reminds oneself that the whole significance of the term "prediction" resides in the temporal relation of its utterance to the event it mentions. "Prediction" is a term defining a category of sentences, in the same way as "command," "argument," "description," etc.; it defines descriptive sentences in the future tense made when the tense is appropriate. The sentence uttered or written in making any given prediction can be repeated after the event has occurred as a perfectly good historical description, provided only that the tense of the verbs (or the corresponding construction) is changed, that is, *apart from tense*, predictions are not identifiable. So this extension of meaning amounts to an *elimination* of the meaning with which one began. We may agree that one procedure of inferring past events is essentially the same as one procedure of inferring future events, but we cannot possibly conclude that the *results* in the first case are essentially the same as predictions. This is like saying that analytic statements are essentially the same as synthetic statements since both can be inferred syllogistically. If the *only way* of inferring to such different kinds of statement was syllogistic, one *might* be more inclined to call them logically identical. One would still not be very impressed, since their logical character is written on their face, and appealing to their common ancestry cannot prove that all siblings are twins; it cannot eliminate the obvious differences. But it is clear that there is nothing unique about the type of inference suggested here; predictions and postdictions can be obtained from arguments of virtually any logical form and also without any argument at all, as in the case of the expert but inarticulate diagnostician or the precognitive. I conclude that the explanation of laws has no proper counterpart among predictions, since there is no general concept of predicting laws; for (i) if what can be predicted is said to be the *discovery* of a law, this fails because the counterpart to explaining an *event* is predicting it, not its *discovery* (which would require laws about discovering laws); (ii) in the only other possible interpretation, a large number of the conclusions inferred are simply not predictions at all; (iii) even ignoring the first two points, nothing is more obvious than the difference in logical structure between the "prediction" "All A's are (or even, will be found to be) C's" and anything that might conceivably be said to be an explanation of it.

(3.45) I think little can be salvaged from the impact of this set of four points (3.41 through 3.44) against the 3.4 thesis, but I wish to indicate another series of difficulties which will help us to develop a constructive alternative position. The first involves a rather lengthy example, but the same example is of some assistance in dealing with the notion of cause as well as those of explanation and prediction. Suppose we are in a position to explain the collapse of a bridge as due to the fatigue of the metal in one of the thrust members. This is not an unusual kind of situation, and it is, of course, one where no prediction was *in fact* made. According to Hempel and Oppenheim, if this is a satisfactory explanation, then, if taken account of in time, it could have formed the basis for a prediction. (We can abandon the idea— presumably but incorrectly taken to be equivalent by Hempel and Oppenheim—that it would actually have the same logical structure as the prediction.) Let us examine in a little detail how this could be so.

We begin our search for the explanation with an eyewitness account that locates a particular girder as the first to go. We already know that there is a substantial deterioration in the elastic properties of carbon steel as it ages and is subjected to repeated compressions; we also know that the amount of this deterioration is not predictable with great reliability since it depends on the conditions in the original welding, casting, and annealing processes, the size and frequency of subsequent temperature changes to which the formed metal is subjected, the special stress to which it may have been subjected, for example, by lightning discharges which put heavy currents through it, and, of course, irregularities in and perhaps violations of the design load. The only way to deal with these sources of error is to "overdesign," that is, to make an allowance for the unpredictables and provide a safety factor on top of that. But the cost of materials and the pressure of competitive tenders puts limits on the size of such safety margins, and every now and then, as in the spectacular case of the Launceston Bridge, where the wind set up resonant vibrations, a failure occurs. In the present case, where internal rather than external circumstances are the significant factor in the failure, we obtain samples of the metal from the girder in question and discover that its elastic properties have substantially deteriorated. But as we do not have any exact data about the load at the time of failure, we cannot immediately prove that such a load would definitely have produced failure.

Now we go over the rest of the bridge carefully, searching for other possible causes of failure and find none. The bridge is of standard design, sited on good bedrock, and well built. We do have good reason to suppose that the load-causing failure was no greater than the bridge had withstood on many previous occasions, though greater than the static load (assume standard traffic and moderate wind); so we are forced to look for the cause in the structural changes. In the light of all this information, we can have great confidence in our explanation of the failure as due to fatigue in the particular beam; but we simply do not have the data required for a prediction that the failure would take place on a certain date.

It is perfectly true that *if* we *also* had exact data about the load when the failure occurred, could obtain some exact and reliable elastic coefficients from the fatigued sample, were in no doubt about our theory, and found on calculation with the revised elastic coefficients that the load exceeded the residual strength, we could be *even more confident* of our explanation. But I have described a much more realistic situation in which we can still have a very high level of rational confidence in our explanation, a level which places it beyond reasonable doubt.

Now, in both cases—with and without the exact details—we *can* make some kind of a *conditional* prediction—that the bridge will fail *if* the load goes over a certain point, for obviously we can give some load which exceeds any known bridge's capacity. It must be noticed first that such a prediction has no practical interest at all except insofar as we can predict the occurrence of such loads. It is a conditional prediction not a categorical prediction, and if the only kind of prediction which is associated with explanations is conditional prediction, especially if they are of this "upper limit" kind, this is of very little interest indeed for scientists or engineers who cannot predict when the conditions are met, or who know they are very rarely met.

These considerations make us realize that the crucial element in the "duality thesis" about explanations and predictions is the existence of a specific correlative prediction for each explanation. Naturally, we can make a number of conditional predictions as soon as, or indeed *before*, we have any data about the material and form of the bridge; but these are independent of the particular circumstances of the failure. The "duality" claim presumably implies that to every different good explanation there corresponds a different prediction relating to precisely those circumstances to which the explanation applies. But it is easy enough to see that we can attain all reasonable certainty about an explanation with less evidence than is required to justify even a conditional prediction with the same *specific* reference.

In the bridge example, we have so far been much too profligate of our investigator's time. In fact, he knows very well that the only causes of failure other than a load in excess of anything for which the structure was originally designed are metal fatigue or external damage by, for example, corrosion, abrasion, or explosion. It is easy to check for the symptoms of external damage, it is relatively easy to judge that the load was not beyond the design limits. Consequently, he can almost certainly identify the cause of failure immediately as fatigue. In suitable circumstances, that is, with suitable evidence for the above statements, it is only a formality to go through with testing samples of the structural steel. Suppose, however, as a final check, we do a rough computation of Young's modulus for material in the beam and find it has substantially decreased and by much more than for the other beams; but we take no exact measurements of it, and none at all of the other elastic coefficients (which normally vary in the same direction as Young's modulus). Our hypothesized explanation has been very strongly confirmed. It is now beyond reasonable doubt. Now what conditional prediction can we make? It seems there is only the extremely weak one that if sufficient substantial fatiguing takes place, and a somewhat higher load than normal is imposed, failure will take place.[11] Not only does such a "prediction" correspond to an indefinitely large number of explanations and hence fail to meet the uniqueness condition previously mentioned, but it is couched in such vague terms as to be almost wholly uninformative. Yet I wish to maintain that such a prediction is all that can be said to be correlated with some very well-established explanations.

Let us examine the most natural counterargument. This would consist in saying that no such explanation could ever be regarded as certain in view of its lack of precise support. Imagine an attorney for the steel company attacking this explanation in court. "How can you be sure that the metal fatigue was *enough* to produce a failure? You made no calculation and no measurements from which you could in any way infer that a bridge built of the same steel, in the same condition you found it, would fail even under *twice* the load impressed on it that windy night. Hence, I submit that no evidence for blaming the steel has been produced, and hence no evidence that the steel was at fault."

The weakness of this argument as far as *our* considerations are concerned is twofold. First, there is not the least difference between direct and indirect evidence for establishing a conclusion beyond reasonable doubt; indirect evidence is often more reliable and the distinction between the two is largely arbitrary. If only A, B, or C can cause X, and A and B are ruled out, it is unnecessary to show C is present; in this case, however, it was *also* shown that C was present, and the only debate is over whether *sufficient* C was present. The reply to such doubts is simply "What else, then?" A redoubtable prima-facie case has been made, and if not rebutted, it must be accepted.

Second, there are some grounds for doubting the significance of any "direct" test, which do not apply to the indirect evidence. Suppose we take all the exact measurements we can and make all the calculations we can, and they indicate that a bridge made of the metal tested would not collapse under any stress that seems likely to have occurred in the circumstances at the time of failure. Here we have two *conflicting* indications. Far from it being the case that the "direct" test affords the crucial test, we find it substantially less reliable than the other evidence. First the sample of metal tested is not known to be identical with that which failed: we take a sample adjacent to the *fracture*, but it is a very difficult matter to determine where the fracture begins, that is, where the *failure* occurs. It is quite certain that different spots on the same girder—and along the same fracture—are under very different stresses and hence at very different stages of fatigue. Second, the steps involved in going from the data on materials to conclusions about bridge strength involve a vast number of assumptions of various kinds, few of them more than approximations whose errors may in sum be fatal to the argument. For example, it is *possible* that exceptional conditions did prevail in local areas around the bridge structure, producing strains such as would not normally be associated with a moderate storm—a typical example is provided by the random development of wind resonance, which can build up a considerable, though not precisely known, extra force from a mild breeze that happens to be blowing in the right direction at the right velocity. Hence our "direct" calculations by no means settle the matter; and the recent examples of wing failure in the Electra airliner show that fatigue can be identified as the cause of failure even when exact theory is wholly inapplicable. The

moral of this example is that explanations can be supported by assertions about qualitative *necessary* conditions whereas even a conditional prediction requires quantitative *sufficient* conditions. (This point would of course be completely lost if one
proceeds on the common assumption that causes are simply sufficient conditions.)

We have thus discovered that the "direct test" of the indirectly supported hypothesis is by no means immune to rejection. But the general issues about confirmation are not important here; it is the existence of *some* cases where we can have
every confidence in an explanation and yet be in no position to make a prediction,
even an applicable conditional prediction. This counterexample to the "duality" view
is the analogue of the counterexamples already mentioned where we are in an excellent position to make a prediction but cannot produce an explanation. A simple
and somewhat rough way of putting the point of the last example would be to say
that a prediction has to say *when* something will happen, or *what* will (sometime)
happen, a causal explanation only *what made it* happen. The first requires either
attaching a time or range of times (unconditional prediction), or a value of some
other variable (conditional prediction), to the description of an event, whereas the
second often requires only giving a cause, that is, picking out (not estimating the
size of) a variable, or another event.

Naturally there are some cases where more than these minimum requirements
are available. Sometimes the nature of the problem is such that when the explanation
is certain the prediction *is* possible: the Farnborough research into the fuselage fatigue of the De Havilland Comet airliners is a case in point. This was possible only
because they had excellent data on the circumstances of the failure (from service
records plus recovered instruments). The first type of case we have described is of
central importance in the social sciences, because most of our knowledge of human
behavior can be expressed only in necessary condition propositions or judgment
propositions. Hence it enables us to explain but not predict with equal accuracy. We
can confidently explain the migration of the Okies to California in terms of the drought
in Oklahoma, though we could not have predicted it with any reliability. For we
know (i) that there must have been a reason for migrating and (ii) that drought produces economic conditions which can provide such a reason, and (iii) that nothing
else with such effects was present. But we do not know *how* much of a drought is
required to produce a migration and hence could not have predicted this with any
confidence. Hempel mistakenly regards this as grounds for doubting the explanation.[12] We must insist on making a distinction between a dubious explanation and
one for which further confirmation—in the technical sense—is still possible: every
empirical claim has the latter property.

(3.46) To summarize, in part. The idea that a causal explanation can only be
justified by direct test of the conditions from which a prediction could be made is
a root notion in the Hempel and Oppenheim treatment of explanation, and they try
to give it a precise formulation. It is said that an explanation must have the form of
a deduction from (a) causal laws (L_v) connecting certain antecedent conditions (C_v)
to certain consequent conditions (E_v), plus (b) assertions that the conditions C_v obtained in the case under consideration, where we are trying to explain X, which is
the sum of the conditions E_v. In the bridge example, we would have to show (by
appeal to connections involving L_v) that material with the properties of the sample
taken ($C_1, C_2, \ldots, C_{n-1}$) under the ambient conditions of the failure ($C_n, C_{n+1}, \ldots,$
C_{m-1}) would lead to the behavior described, that is, X, the collapse of the bridge

(the bridge's design and state prior to failure being described in terms of C_m, C_{m+1}, ..., C_p). I have been arguing that an indirect approach may be just as effective, that is, one showing fatigue (C_1) to be a necessary condition for X under the circumstances (C_2, ..., C_p). This would involve appealing to a proposition of the form "If X occurs, then either A_1, B_1, or C_1 caused it" and showing that C_2, ..., C_p rule out A_1, and B_1. The main trouble with such laws for the thesis of Hempel and Oppenheim is that they do not permit any predictions of X, since the occurrence of X is required for their application. Nor can such laws be reformulated for predictive use, for they are quite different from "If A_1 or B_1 or C_1 occur, they will produce X," not just because one states necessary and the other sufficient conditions, but because the first does not and the second does require quantitative formulation if it is to be true— for it is obviously false that *any* degree of fatigue produces failure. These laws incidentally demonstrate that the duality thesis about explanation and prediction was actually a separate, fifth condition and not a consequence of the four conditions R_1–R_4.

(3.47) In concluding this discussion of the prediction criterion for explanations, I think it is worth mentioning some points which are neither wholly independent of those discussed above nor, it seems to me, quite so strong. First, it is a consequence of Hempel and Oppenheim's analysis that whatever we explain must be a true statement, since they explicitly require all statements in an explanation to be true. Now it is certainly not the case that all the predictions we make must be true: we often err in predicting the behavior of the stock market, the weather, and the ponies. This point is thought by Scheffler to show that explanations and predictions are different in this respect; but I take it to be mainly a difficulty with Hempel and Oppenheim's analysis of explanation. For one can talk of explaining things that do not happen, just as one can talk of the consequences of things that do not happen. "If you hadn't got here on time, I know who would have been responsible," the irate parent says to the almost-wayward daughter; "If the fourth stage had failed to (ever fails to) fire, you may be sure it would have been (will be) because of a valve failure in the fuel-line," the missile technician may say. This use is *derivative*, that is, it can be explained by reference to the commonest use; and in the commonest use I think we can agree that to say something is an explanation of X is to presuppose (in Strawson's sense) that X occurred. But this is not to say that in *all proper uses*, this can be inferred.

Apart from the case just cited, where it is known that X did not occur, known even by the giver of the explanation, and apart from explanations of events in fiction, there are other cases where this condition does not hold. In the modified phlogiston theory of about 1785, the explanation of the limited phlogistication of air when calcination occurs in a closed vessel was in terms of the finite capacity of a finite volume of air for absorbing phlogiston. The very phenomenon here explained does not exist, although even Cavendish thought it did. The explanation given is within the theory, of something described in *theoretical* terms. This is not to be confused with the case where we quite commonly put single quotation marks around the term "explanation," meaning that the term is not properly applicable, as when Conant says "an 'explanation' of metallurgy was at hand: Metallic ore (an oxide) + Phlogiston from charcoal → Metal."[13] For here we are referring to an incorrect explanation of something we know *does* occur, viz. smelting, and we know this not to be the correct explanation: compare the previous cases discussed where we know the phenomenon *does not* occur. I am therefore unwilling to agree that all proper uses of the term "explanation" presuppose that the phenomenon explained occurs; though I would agree that in the primary use this is so.[14] I shall say something about the necessity for the truth of the body of the explanation itself in the next section.

In the primary use of explanation, then, we know something when we are called on for an explanation that we do not know when called on for a prediction, viz. that the event referred to has occurred. This is sometimes a priceless item of information, since it may demonstrate the existence or absence of a hitherto unknown strength of a certain power. Thus, to take a simpler example than the bridge case, a man in charge of an open-hearth furnace may be suspiciously watching a roil on the surface of the liquid steel, wondering if it is a sign of a "boil" (an occasionally serious destructive reaction) on the furnace lining down below or just due to some normal oxidizing of the additives in the mixture. Suddenly, a catastrophe: the whole charge drops through the furnace lining into the basement. It is now absolutely clear that there was a boil which had eaten through the lining: apart from sabotage (easily disproved by examination) there's no other possibility. But no prediction is possible to the event, using the data then available. This renders almost empty Hempel and Oppenheim's (and even Scheffler's) conclusion that explanations provide a basis for predictions. For "Had we known what was going to happen, we could have predicted it" is a vacuous claim. One might mutter something about "If the furnace was in exactly the same state *again* we could predict it would dump," but I have already pointed out that this is a virtually empty remark since we usually can't identify "exactly the same state"; it is simply a dubious determinist slogan, not even a genuine conditional prediction. Since it is technically entirely impossible to rebuild the furnace to the point where it is identical down to the temperature distribution in the mixture (a crucial factor) and the shape of the irregularities in the floor (also crucial), even if we knew these specifications, it will be pure chance if the conditions ever recur and when they do they won't be identifiable. Thus our grounds for thinking the determinist's slogan to be true—if we do—are entirely indirect, and the explanation certainly does not rest on subsumption under the slogan since we cannot even tell when the latter applies, whereas we can be sure the explanation is correct.

The problem of direct versus indirect confirmation which arises here is of great importance throughout structural logic. To say that "same cause, same effect" is a determinist's slogan is not to say it has *no* empirical content. It has, and it is actually false, as far as present evidence goes, though only to a small extent for macroscopic observations. (It is also not equivalent to the idea of determinism as universal law-governed behavior.) What it lacks is single-case applicability and hence direct confirmability when complex systems are involved—for it is often impossible to specify what counts as the "same conditions." It may still be felt on general grounds that unless we do know what counts as the "same conditions" in a given case, we cannot be sure of the proposed causal explanation in that case. The opposite thesis will be defended in a later section of this paper, and to prepare for it I shall need to make several further distinctions and points. At this stage, however, let me summarize by saying that any prediction specifically associated with an explanation is (i) often conditional, and (ii) either so general as to be almost empty or so specific as to refer to no other case, and (iii) often not assertible until it is known the event occurred, that is, not a true prediction.

3.5. Explanations as Sets of True Statements

It is not possible to claim that explanations can only be offered for events that actually occur or have occurred. They can be given for events in the future (Scheffler), for events in fiction, for events known not to occur, and for events wrongly believed to occur—and also for some laws, states, and relationships which are timeless (see above). Assuming Hempel and Oppenheim's analysis to be in other respects correct, it follows that in such cases some of the propositions comprising the explanation itself cannot be true, contrary to one of their explicit conditions.[15] The reason they give for this condition is a very plausible one, however, and it is of interest to see if a more general account can be given which will contain allowance for their point. They say: ". . . it might seem more appropriate to stipulate that the (explanation) has to be highly confirmed by all the relevant evidence rather than that it should be true. This stipulation, however, leads to awkward consequences. Suppose that a certain phenomenon was explained at an earlier stage of science by means of an (explanation) which was well supported by the evidence then at hand, but which had been highly disconfirmed by more recent empirical findings. In such a case, we would have to say that originally the explanatory account was a correct explanation, but that it ceased to be one later, when unfavorable evidence was discovered. This does not appear to accord with sound common usage, which directs us to say that . . . the account in question was not—and had never been—a correct explanation.[16]

It is roughly on these grounds that Conant puts the term "explanation" in quotes when he is referring to the phlogiston theory's account of calcination. For much the same reason we refer to an astrologer's remarks as an "explanation" of Henry Ford's successful business career.

But notice we can talk perfectly well about "two competing explanations" of some phenomenon in contemporary physics without feeling it improper to refer to both as explanations although only one can be true. And there certainly seem to be cases where we want to say, for example, that the Babylonian explanation of the origin of the universe was basically naturalistic, without using inverted commas. The best treatment of these cases, it seems to me, is to regard them as secondary uses which have become fairly standard, the notion of a secondary use being defined in terms of the fact that understanding it depends logically on understanding the primary use. But these are definitely proper uses and the term "explanation" is hence perhaps less a 'success word' or 'achievement word' than, for example, "knowledge" and "perception." We canot say of two contradictory claims that both are *known*, since this implies both are true. And this suggests a solution to our present problem.

The proper way of avoiding Hempel and Oppenheim's powerful argument is, I think, very simple; the secondary uses of "explanation" are legitimate but there are no such secondary uses of "correct explanation," the term which they substitute halfway through the argument. Remove the qualifying adjective "correct" and you will see that the argument is no longer persuasive. For consistency, this term must be and can be added to the occurrences of "explanation" in the premises. Overwhelming counterevidence does not necessarily lead us to abandon or even to put quotes around "explanation," but, as the argument rightly says, it does lead us to abandon the application of the term "correct explanation" (or "the explanation" which is often used equivalently). Hence we should regard Hempel and Oppenheim's analysis as an analysis of "correct explanation" rather than of "explanation," or "an explanation," and this is surely what they were most interested in. "Explanations," or "an explanation," or "his explanation," or "a possible explanation," do not *always* have to be true (or of the appropriate type, or adequate); they only need high confirmation, at some stage.

Doesn't the notion of confirmation come into the analysis of "correct explanation" at all? It is not part of the *analysis,* which only involves truth; but it is our only *means of access* to the truth. We have not got the correct explanation unless it contains only true assertions, but if we want to know which explanation is most likely to meet that condition, we must select the one with the highest degree of confirmation. Good evidence does not guarantee true conclusions but it is the best indicator, so we need no excuse for appealing to degree of confirmation. Moreover, we have no need to adopt the skeptic's position that all possibility of knowing when we have a correct explanation is by now beyond reasonable doubt, and to restrict "knowing" to cases of absolute logical necessity is to mistake the empty glitter of definitional truth for the fallible flame of knowledge. The notion of reasonable doubt is highly dependent on context, but highly unambiguous in a given context, and it sets the threshold level which distinguishes knowledge from likelihood. Anything that is to be called the "correct explanation" of something that is known to have happened must contain only statements from the domain of knowledge.

Now among the things we know are some statements about the probability of certain events under certain circumstances, for example, about the probability of throwing a six with a die that passes various specifiable tests. Could we not use such propositions as part of an explanation? Hempel and Oppenheim—in the papers cited—countenance the possibility that what they call statistical explanations may be of great importance but they neither undertake to discuss them nor, more significantly, restrict most of their conclusions about explanation in general in the way that would be appropriate if we do take seriously the claims of statistical explanations (which I include in the broader class of explanations based on probability statements). In *Minnesota Studies*, vol. 3, Hempel sets out an account of statistical explanation on which I comment later. Such explanations cannot be subsumed under Hempel and Oppenheim's original analysis as it stands, because no *deduction* of a nonprobability statement from them is possible, and it is hence impossible for them to explain any actual occurrence, since actual occurrences have to be described by non-probability statements. In particular we could make a (probable) prediction from such "laws," but could not—using the same premises—be said to explain the event predicted, if it does come about.

(3.51) It is of some importance to notice that Hempel and Oppenheim's analysis of explanation absolutely presupposes a descriptive language. For them there can be nothing to explain if there is no language, since the thing-to-be-explained is dealt with *only* via the "explanandum" which is its description in the relevant language. One suspects such a restriction immediately because there are clearly cases where we can explain without language, for example, when we explain to the mechanic in a Yugoslav garage what has gone wrong with the car. Now this is hardly a scientific explanation, but it seems reasonable to suppose that the scientific explanation represents a refinement on, rather than a totally different kind of entity from, the ordinary explanation. In our terms, it is the *understanding* which is the essential part of an explanation and the *language* which is a useful accessory for the process of communicating the understanding. By completely eliminating consideration of the step from the phenomenon to the description of the phenomenon, Hempel and Oppenheim make it much easier to convince us that deducibility is a criterion of explanation. In fact, within the language there is only one other relation possible, viz. inducibility. We shall argue that good inductive inferribility is the only required relation involved in explanations, deduction being a dispensable and overrestrictive requirement which may of course sometimes be met. But a source of both error and understanding is left out of account in such a debate; for unexplained things are sometimes such that we do not describe them in asking for an explanation and such that they are explained *merely* by being described in the correct way regardless of deduction from laws.[17] (And on the other hand, sometimes a *true* description and deduction is not enough.)

3.6. Explanations as Involving Descriptions of What Is to Be Explained

Once we have realized the extraordinary difficulty there is in supposing that explanations and predictions have a common structure, it is natural to ask what the structure of an explanation really is. The "structure" of a prediction, we noticed, is simply that of a declarative statement using an appropriate future tense and any kind of descriptive language. It *may* indeed be of the form "*C* will bring about *X*" but is more usually of the form "*X* will occur" or "*X*, at time *t*." The structure of an argument, to take a further example, is such as to involve several statements which are put forward as bearing upon each other or upon some other statements in the relation of premises to conclusions. Now what is the structure of an explanation? A bridge's failure may have as its explanation the fatigue of the metal in a particular member or the overload due to a bomb blast. These appear to be a state and an event which could be held to be the cause of the event to be explained. A different account will have to be given of the explanation of laws, but for the moment we can profitably concentrate on the explanation of events, to which Hempel and Oppenheim devote a good deal of space, and which has some claim to be epistemologically prior to the explanation of laws.

Now there is a further apparent ambiguity about "explanation": it can either refer to the linguistic structure which describes certain states or events or to the states (or events) themselves. This kind of ambiguity even occurs in connection with such terms as "consequence," "concept," "cause," "inference," and "argument"; it is common throughout logic and best illustrated, perhaps, by the very term "fact." We shall usually be referring to the linguistic entity when we use the term "explanation," but clearly *neither* this entity nor its referents include whatever it is that is to be explained. In the simple but standard examples just given, the explanation, in this sense, is an assertion about a state or event that is entirely different from (assertions about) the state or event to be explained. But Hempel and Oppenheim say, "We divide an explanation into two major constituents . . . the sentence describing the phenomenon to be explained . . . [and] the class of those sentences which are adduced to account for the phenomenon."[18] The former is plainly not a constituent of the explanation at all (except where it is all of the explanation—see 3.5). Only if we find its consequences very confusing or inconvenient should we abandon such a clear distinction as this.

The first difficulty that strikes us about this version of the Hempel and Oppenheim account, then, is that it asserts all explanations of a phenomenon X consist in a deductive argumentlike structure, with a statement about X as the conclusion, whereas our simple examples above are merely statements about something or other that is held to be the cause of whatever is to be explained. And are there not occasions, on which one is going over—demonstrating—an explanation, when one does finish off by giving as the last step the description of what is to be explained? It is certainly not a common practice, scientifically or ordinarily, and even when it occurs, it only shows that part of a proof that something is an explanation of X may involve a description of X. The *explanation* of the photoelectric effect does not involve the description of the effect—this is presupposed by the explanation. The point may be minor, but it puts us on our guard, for we cannot be sure whether it may not have unfortunate consequences, analogous to those involved in saying that predictions have the same logical structure as explanations. In fact, we have already seen one error that results from this incautious amalgamation of (i) phenomena, (ii) their description, and (iii) their explanation, in 3.5. We could state part of it by saying that a sixth requirement is actually implicit in their account, viz. the requirement of accuracy and relevance of the description of X, which for them is part of the explanation of X.

In fact, the most serious error of all those I believe to be involved in Hempel and Oppenheim's analysis also springs from the very same innocuous-seeming oversimplification: the requirement of deducibility itself, plausible only if we forget that our concern is fundamentally with a phenomenon, not a statement. It may seem unjust to suggest that Hempel and Oppenheim amalgamate the phenomenon and its description (though certainly they do amalgamate the description and the explanation) when they make clear that the "conclusion" of the explanation is "a sentence describing the phenomenon (not that phenomenon itself)."[19] The justice of my complaint rests, not on their failure *ever* to make this distinction, but on their failure to be consistent in dealing with its consequences. For it is a consequence of this distinction that a nondeductive step is involved between the statements in an explanation and the phenomenon explained. And we may then ask why they should suppose deducibility to be the only logical relation in a good explanation. They never address themselves to this question directly, chiefly, I think, because they do not *realize* the consequences of the distinction they do once make. Attention to it would surely have led them to notice (i) cases of explanatory description (see 3.5), (ii) cases where the completeness or (iii) the uniqueness of the description are crucial in assessing the explanation (see 6.2). Only if we assume that getting as far as the description is getting to the phenomenon, that is, doing what an explanation is supposed to do, could we overlook such interesting cases. (I think the fact that "description" can be taken to *mean* "accurate description" also led them to overlook the independent importance of this requirement.)

3.7. The Last Two Conditions and a Summary of Difficulties

It is stated that the explanation "must contain general laws, and these must actually be required for the derivation. . . ." And finally, it is said that the derivation must be deductive, ". . . for otherwise the (explanation) would not constitute adequate grounds for (the proposition describing the phenomenon)."[20] We now have a general idea of Hempel and Oppenheim's model of explanation, which I have elsewhere christened, for obvious reasons, 'the deductive model.'[21] I wish to maintain against it the following criticisms in particular, and some others incidentally;

> 1. It fails to make the crucial logical distinctions between explanations, grounds for explanations, predictions, things to be explained, and the description of these things.
> 2. It is too restrictive in that it excludes their own examples and almost every ordinary scientific one.
> 3. It is too inclusive and admits entirely nonexplanatory schema.
> 4. It requires an account of cause, law, and probability which are basically unsound.
> 5. It leaves out of account three notions that are in fact essential for an account of scientific explanation: context, judgment, and understanding.

These objections are not wholly independent, and I have already dealt with some of them.

4. Fundamental Issues

4.1. The Distinction Between Explanations and the Grounds for Explanations

It is certainly not the case that our grounds for thinking a plain descriptive statement to be true are part of the statement itself; no one thinks that a more complete analysis of "Gandhi died at an assassin's hand in 1953" would include "I read about Gandhi's death in a somewhat unreliable newspaper" or "I was there at the time and saw it happen, the only time I've been there, and it was my last sabbatical leave so I couldn't be mistaken about the date," etc. Why, then, should one suppose that our grounds for (believing ourselves justified in putting forward)[22] a particular explanation of a bridge collapsing, for example, the results of our tests on samples of the metal, our knowledge about the behavior of metals, eyewitness accounts, are part of the explanation? They might indeed be produced as part of a *justification* of (the claim that what has been produced is) the explanation. But surely an explanation does not have to contain its own justification any more than a statement about Gandhi's death has to contain the evidence on which it is based. Yet, the deductive model of explanation requires that an explanation include what are often nothing but the grounds for the explanation.

Not only linguistic impropriety but absolute impossibility is involved in the attempt to market the joint package as the "whole explanation" or "complete explanation." The linguistic impropriety is twofold: first, perfectly proper explanations would be rejected for the quite unjust reason that they did not contain the grounds on which they were asserted; second, the indefinite number of possible grounds for an explanation makes absurd the idea of a single correct explanation since there is, in terms of the model, nothing more or less correct about any one of the wide range of possible sets of deductively adequate true grounds. And clearly these are circumstances in which we do identify a particular account as "The correct explanation." The impossibility derives from the second impropriety. There is no sense in which one could ever provide a complete justification of an explanation, out of context;

for a justification is a defense against some specific doubt or complaint, and there is an indefinite number of possible doubts.

The deductive model apparently provides an answer to the latter objection in an interesting way. It prescribes that the only kind of justification required is deduction from general laws and specific antecedent conditions. Once this is given, a complete explanation has been given; until this has been done, only (at best) an "explanation sketch" has been given.

When we say that a perfectly good explanation of one event, for example, a bridge collapsing, may be no more than an assertion about another event, for example, a bomb exploding, might it not plausibly be said that this can only *be* an explanation if some laws are assumed to be true, which *connect* the two events? After all, the one is an explanation of the other, not because it came before it, but because it *caused* it. In which case, a full statement of the explanation would make explicit these essential, presupposed laws.

The major weakness in this argument is the last sentence; we can put the difficulty again by saying that, if completeness requires not merely the existence but the quoting of all necessary grounds, there are no complete explanations at all. For just as the statement about the bomb couldn't be an explanation of the bridge collapsing unless there was some connection between the two events, it couldn't be an explanation unless it was true. So, *if* we must include a statement of the relevant laws to justify our belief in the connection, that is, in the soundness of the explanation, then we must include a statement of the relevant data to justify our belief in the claim that a bomb burst, on which the soundness of the explanation also depends.[23]

Certainly in putting forward one event as an explanation of another in the usual cause-seeking contexts, we are committed to the view that the first event caused the second, and we are also committed to the view that the first took place. Of course, we may be wrong about either view and then we are wrong in thinking we have given the explanation. But it is a mistake to suppose this error can be eliminated by quoting further evidence (whether laws or data); it is merely that the error may be then located in a more precise way—as due to a mistaken belief in such and such a datum or law. The function of deduction is only to shift the grounds for doubt, though doubts sometimes get tired and give up after a certain amount of this treatment.

Perhaps the most important reason that Hempel and Oppenheim have for insisting on the inclusion of laws in the explanation is what I take to be their belief (at the time of writing the paper in question) that only if one had such laws in mind could one have any rational grounds for putting forward one's explanation. This is simply false as can be seen immediately by considering an example of a simple physical explanation of which we can be quite certain. If you reach for a cigarette and in doing so knock over an ink bottle which then spills onto the floor, you are in an excellent position to explain to your wife how that stain appeared on the carpet, that is, why the carpet is stained (if you cannot clean it off fast enough). You knocked the ink bottle over. This is the explanation of the state of affairs in question, and there is no nonsense about it being in doubt because you cannot quote the laws that are involved, Newton's and all the others; in fact, it appears one cannot here quote any unambiguous true general statements, such as would meet the requirements of the deductive model.

The fact you cannot quote them does not show they are not somehow *involved*, but the catch lies in the term "involved." Some kind of connection must hold, and if we say this *means* that laws are involved, then of course the point is won. The suggestion is debatable, but even if true, it does not follow that we will be able to state a law that guarantees the connection. The explanation requires that there be a connection, but not any particular one—just one of a wide range of alternatives. Certainly it would not be the explanation if the world was governed by *anti*gravity. But then it would not be the explanation if you *had not* knocked over the ink bottle— and you have just as good reasons for believing that you did knock it over as you have for believing that knocking it over led to (caused) the stain. Having reasons for causal claims thus does not always mean being able to quote laws. We shall return to this example later. For the moment, it is useful mainly to indicate that (i) there is a reply to the claim that one cannot have good reasons for a causal ascription unless one can quote intersubjectively verifiable general statements and (ii) there is an important similarity between the way in which the production of an appropriate law supports the claim that one event explains another, and the way in which the production of further data (plus laws) to confirm the claim that the prior event occurred supports the same claim. They are defenses against two entirely different kinds of error or doubt, indeed, but they are also both support for the same kind of claim, viz. the claim that one event (state, etc.) explains another.

This is perhaps obscured by the fact that when we make an assertion our claim is in full view, so to speak, whereas when we put forward an assertion *as* an explanation, its further role is entirely derived from the context, for example, that it is produced in answer to a request for an explanation, and so its further obligations seem to require explicit statement. This is a superficial view. All that we actually identify in the linguistic entity of a 'declarative statement' is the subject, predicate, tense, etc. We have no reason at all, apart from the context of its utterance, for supposing it to be *asserted,* rather than proposed for consideration, pronounced for a grammatical exercise, mouthed by an actor, produced as an absurdity, etc.[24] *That* it is asserted to be true we infer from the context just as we infer that it is proffered as the explanation of something else; and for both these tasks it may need support. We may concede that assertion is the *primary* role of indicative sentences without weakening this point.

It is in fact the case that considerations of context, seen to be necessary even at the level of identifying assertions and explanations themselves, not only open up another dimension of error for an explanation, that of pragmatic inappropriateness, but simultaneously offer a possible way of *identifying* the explanation of something, where this notion is applicable.

A particular context—such as a discussion between organic chemists working on the same problem—may make one of many deductively acceptable explanations of a biochemical phenomenon entirely inappropriate, and make another of exactly the right type. (Of course, I also wish to reject the criteria of the deductive model; but even if one accepted it, the consideration of context turns out to be *also* necessary. So its importance is not only apparent in dealing with alternative analyses.)

We may generalize our observations in the following terms. An explanation is sometimes said to be incorrect or incomplete or improper. I suggest we pin down these somewhat general terms along with their slightly more specific siblings as follows. If an explanation explicitly contains false propositions, we can call it *incorrect* or *inaccurate*. If it fails to explain what it is supposed to explain because it cannot be "brought to bear" on it, for example, because no causal connection exists between the phenomenon as so far specified and its alleged effect, we can call it *incomplete* or *inadequate*. If it is satisfactory in the previous respects but is clearly not the explanation required in the given context, either because of its difficulty or its field of reference, we can call it *irrelevant, improper,* or *inappropriate*.

Corresponding to these possible failings there are types of defense which may be relevant. Against the charge of inaccuracy, we produce what I shall call *truth-justifying grounds*. Against the charge of inadequacy, we produce *role-justifying grounds,* and against the complaint of inappropriateness, we invoke *type-justifying grounds*. To put forward an explanation is to commit oneself on truth, role, and type, though it is certainly not to have explicitly considered grounds of these kinds in advance, any more than to speak English in England implies language-type consideration for a lifelong but polylingual resident Englishman.

The mere production of, for example, truth-justifying grounds does not guarantee their acceptance, of course. They may be questioned, and they may be defended further by appeal to further evidence; we defend our claim that a bomb damaged a bridge by producing witnesses or even photographs taken at the time; and we may defend the accuracy of the latter by producing the unretouched negatives and so on. The second line of defense involves *second-level grounds,* and they may be of the same three kinds. That they can be of these kinds is partly fortuitous (since they are not explanations of anything) and due to the fact that the relation of being-evidence-for is in certain ways logically similar to being-an-explanation-of. In each case, truth, role, and type may be in doubt; in fact, this coincidence of logical character is extremely important. We notice, however, that there is no similarity of any importance between these two and being-a-prediction-of; where truth is not relevant in the same way, role is wholly determined by time of utterance and syntax, and only the type can be—in some sense—challenged.

4.2. Completeness in Explanations

The possibility of indefinitely challenging the successive grounds of an explanation has suggested to some people—*not* Hempel and Oppenheim—that a complete explanation cannot be given within science. Such people are adopting another use of "complete"—even less satisfactory than Hempel and Oppenheim's—according to which the idea of a complete explanation becomes not only foreign to science but in fact either wholly empty, essentially teleological, or capable of completion by appeal to a self-caused cause. Interesting though this move is in certain respects, it essentially requires saying that we can better understand something in the world by ultimately ascribing its existence and nature to the activities of a mysterious entity whose existence and nature cannot be explained in the same way, than by relating it to its proximate causes or arguing that the world has existed indefinitely. I shall only add that we are supposed to be studying scientific explanations, and if none of them are complete in this sense, we may as well drop this sense while making a note of the point—which is equivalent to the point that the causal relation is irreflexive and hence rather unexciting—for there is an important and standard use of "complete" which does apply to some suggested scientific explanation, and not to others, and is well worth analyzing.

Now, if some scientific explanations are complete—and think how a question in a physics exam may ask for a complete explanation of, for example, the effects noticed by Hertz in his experiments to determine whether electromagnetic waves existed—it cannot be because there is a last step in the process of challenging grounds, for there is no stage at which a request for further proof could not make sense. But *in any given context* such requests eventually become absurd, because in any scientific context certain kinds of data are taken as beyond question, and there is no meaning to the notions of explanation and justification which is not, directly or indirectly, dependent on a context. This situation is of a very familiar kind of logic. It makes perfectly good sense to ask for the spatial location of any physical object; and perfectly proper and complete answers will involve a reference to the location of some other physical objects. Naturally we can, and often do, go on to ask the further question concerning where these other objects are ("Where's Carleton College?" "In Northfield." "But where's Northfield?"). And no question in a series of this kind is meaningless, unless one includes the question "Where is the universe, that is, everything?" as of this kind. If one does include this question (which is the analogue of "Where did the universe come *from?*"), then the impossibility of answering it only shows something about the notion of position, and nothing about the incompleteness of our knowledge. If one excludes this question, the absence of a last stage in such a series does not show our inability to give anyone complete directions to the public library but ony that the notion of completeness of such descriptions involves context criteria.

Any request for directions logically presupposes that *some* directions can be understood; if no directions can be understood, then the proper request is for an account of the notions of position and directions. A *complete* answer has been given when the particular object has been comprehensibly related to the directions that are understood. Similarly, then, the request for an explanation presupposes that *something* is understood, and a complete answer is one that relates the object of inquiry to the realm of understanding in some comprehensible and appropriate way. What this way is varies from subject matter to subject matter just as what makes something better than something else varies from the field of automobiles to solutions of chess problems; but the *logical* function of explanation, as of evaluation, is the same in each field. And what counts as complete will vary from context to context within a field; but the logical category of complete explanation can still be characterized in the perfectly general way just given, that is, the logical function of "complete," as applied to "explanations," can be described. Hence the notion of the proper context for giving or requesting an explanation, which presupposes the existence of a certain level of knowledge and understanding on the part of the audience or inquirer, *automatically* entails the possibility of a complete explanation being given. And it indicates exactly what can be meant by the phrase "*the* (complete) explanation." For levels of understanding and interest define areas of lack of understanding and interest, and the required explanation is the one which relates to these areas and not to those other areas related to the subject of the explanation but perfectly well understood or of no interest (these would be explanations which could be correct and adequate but inappropriate). It is worth mentioning that the same analogy with spatial location (or evaluation) provides a resolution of the "problem of induction," as a limit case of a request for a "complete justification."

It is also clear that calling an explanation into question is not the same as—

though it includes—rejecting it as not itself explained. Type justifying involves more than showing relevance of subject matter, that is, topical and ontological relevance; it involves showing the appropriateness of the intellectual and logical level of the content; a proposed explanation may be inappropriate because it involves the wrong kind of true statements from the right field, for example, trivial generalizations of the kind of event to be explained, such that they fulfill the deductive model's requirements but succeed only in generalizing the puzzlement. One cannot explain why this bridge failed in this storm by appealing to a law that all bridges of this design in such sites fail in storms of this strength (there having been only two such cases, but there being independent evidence for the law, not quoted). This might have the desirable effect of making the maintenance boss feel responsible, but it surely does not explain *why* this bridge (or any of the other bridges of the same design) fails in such storms. It may be because of excessive transverse wind pressure, because of the waves affecting the foundations or lower members, because of resonance, etc.

So mere deduction from true general statements is again seen to be less than a sufficient condition for explanation; but what interests us here is that our grounds for *rejecting* such an explanation are not suspicions about its *truth* or its *adequacy,* which are the usual grounds for doubting an explanation, but only its failure to *explain.* Certainly it fails to explain if incorrect or inadequate, but then one feels it fails in a genuine attempt, that the slip is then between the cup and the lip; whereas irrelevance of type is a slip between the hand and the cup—the question of it being a *sound* explanation never even arises. One may react to this situation by declaring with Hempel and Oppenheim that the only *logical* criteria for an explanation are correctness and adequacy, the matter of type being psychological; or, as I think preferable, by saying that the concept of explanation is logically dependent on the concept of understanding, just as the concept of discovery is logically dependent on the concept of knowledge-at-a-particular time. One cannot discover what one already knows, nor what one never knows; nor can one explain what everyone or no one understands. These are tautologies of logical analysis (I hope) and hardly grounds for saying that we are confusing logic with psychology.

Having distinguished the types of difficulty an explanation may encounter, one can more easily see there is no reason for insisting that it is complete only if it is armed against them in advance, since (i) to display in advance one's armor against *all* possible objections is impossible and (ii) the value of such a requirement is adequately retained by requiring that scientific explanations be such that scientifically sound defenses of the several kinds indicated *be available* for them though not necessarily *embodied in* them. Since there is no special reason for thinking that true first-level role-justifying assumptions are any more necessary for the explanation than any others, it seems quite arbitrary to require that they should be included in a complete explanation; and it is quite independently an error to suppose they must take the form of laws.

Notes

1. "The Logic of Explanation" in *Readings in the Philosophy of Science,* H. Feigl and M. Brodbeck, eds. (New York: Appleton-Century-Crofts, 1953), p. 319. This is an abridg-

ment of "Studies in the Logic of Explanation," *Philosophy of Science*, 15, 1948, pp. 135–75. All page references hereafter are to the later version.

2. *Scientific Explanation* (Cambridge: Cambridge University Press, 1953), p. 319.

3. The discussion of How possibly questions has been initiated and sustained by William Dray. See his *Laws and Explanation in History* (London: Oxford University Press, 1957), especially pp. 164ff.

4. Op. cit., p. 319.

5. For an alternative and acceptable interpretation of their remarks, see 3.4 below.

6. Op. cit., p. 320.

7. Op. cit., pp. 322–23.

8. And this salvages the theme which forms the title of 3.3—but at the expense of making that of 3.4 untenable. I am thus uncertain which interpretation to accept.

9. See also "Explanation, Prediction, and Abstraction" by Israel Scheffler, *British Journal for the Philosophy of Science*, 7: 293–309 (1957), where several of the points in this section are discussed.

10. The reader may be worried by the fact that his law is known to be only an approximation. This is true of almost all "laws," but we do give explanations of them, for example, of Kepler's laws, Snell's law. Now, for any such law, deducing it from any premises would simply show the premises to be inaccurate. Hence, explanation cannot require deduction from true premises. We must substitute a weaker requirement; there are several possibilities which appear to retain much of the Hempel analysis (but see 3.5 below).

11. Hempel, replying to this point in [*Minnesota Studies in the Philosophy of Science*, vol. 3], suggests another, more specific candidate. I comment on this later.

12. "The Function of General Laws in History," *Journal of Philosophy*, 39, (1942), 35–48.

13. *Harvard Case Histories in Experimental Science*, vol. I (Cambridge, Mass.: Harvard University Press, 1958), pp. 70, 110.

14. There are of course a number of terms besides "explanation" (for example, description (see 3.6), observation, insight) that are used in such a way that the description "incorrect (explanation)" can be synonymous with "not an (explanation) at all." The points just made do not, however, depend on this ambiguity.

15. Op. cit., pp. 321–22.

16. Op. cit., p. 322.

17. A common case is that when someone, greatly puzzled, asks What on earth is this? or What's going on here? and is told, for example, that it is an initiation ceremonial on which he has stumbled. Analogous cases in particle physics, engineering, and astronomy are obvious. The point of these examples is that understanding is roughly the perception of relationships and hence may be conveyed by any process which locates the puzzling phenomenon in a system of relations. When we supply a law, we supply part of the system; but a description may enable us to supply a whole framework which we already understand, but of whose *relevance* we had been unaware. We deduce nothing; our understanding comes because we *see* the phenomenon for what it is, and are *in a position* to make other inferences from this realization.

18. Op. cit., p. 321.

19. Op. cit., p. 321.

20. Op. cit., p. 321.

21. "Certain Weaknesses in the Deductive Model of Explanation," paper read at the Midwestern Division of the American Philosophical Association, May 1955.

22. I shall abbreviate some more precise formulations by omitting the words in parentheses where I think they are not essential.

23. Their model requires the truth of the asserted explanation, but it doesn't require the

inclusion of evidence for this. Instead of similarly requiring a causal connection, it actually requires the inclusion of one *special kind* of evidence for this. If it treated both requirements equitably, the model would be either trivial (causal explanations must be true and causally relevant) or deviously arbitrary (. . . must include *deductively* adequate grounds for the truth of any assertions and for the causal connection).

24. See Max Black's "Definition, Presupposition, and Assertion," in his *Problems of Analysis* (London: Routledge and Kegan Paul, 1954).

4

Statistical Explanation and Causality

Wesley C. Salmon

1. Statistical Explanation

The Nature of Statistical Explanation

Let me now, at long last, offer a general characterization of explanations of particular events. As I have suggested earlier, we may think of an explanation as an answer to a question of the form, "Why does this x which is a member of A have the property B?" The answer to such a question consists of a partition of the reference class A into a number of subclasses, all of which are homogeneous with respect to B, along with the probabilities of B within each of these subclasses. In addition, we must say which of the members of the partition contains our particular x. More formally, an explanation of the fact that x, a member of A, is a member of B would go as follows:

$$P(A.C_1,B) = p_1$$
$$P(A.C_2,B) = p_2$$
$$\vdots$$
$$P(A.C_n,B) = p_n$$

where

$A.C_1, A.C_2, \ldots, A.C_n$ is a homogeneous partition of A with respect to B,

$p_i = p_j$ only if $i = j$, and

$x \in A.C_k$.

With Hempel, I regard an explanation as a linguistic entity, namely, a set of statements, but unlike him, I do not regard it as an argument. On my view, an explanation is a set of probability statements, qualified by certain provisos, plus a statement specifying the compartment to which the explanandum event belongs.

The question of whether explanations should be regarded as arguments is, I be-

Reprinted from *Statistical Explanation and Statistical Relevance* by Wesley Salmon by permission of the University of Pittsburgh Press. © 1971 by University of Pittsburgh Press.

lieve, closely related to the question, raised by Carnap, of whether inductive logic should be thought to contain rules of acceptance (or detachment).[1] Carnap's problem can be seen most clearly in connection with the famous lottery paradox. If inductive logic contains rules of inference which enable us to draw conclusions from prem- ises—much as in deductive logic—then there is presumably some number r which constitutes a lower bound for acceptance. Accordingly, any hypothesis h whose prob- ability on the total available relevant evidence is greater than or equal to r can be accepted on the basis of that evidence. (Of course, h might subsequently have to be rejected on the basis of further evidence.) The problem is to select an appropriate value for r. It seems that no value is satisfactory, for no matter how large r is, provided it is less than one, we can construct a fair lottery with a sufficient number of tickets to be able to say for each ticket that [it] will not win, because the prob- ability of its not winning is greater than r. From this we can conclude that no ticket will win, which contradicts the stipulation that this is a fair lottery—no lottery can be considered fair if there is *no* winning ticket.

It was an exceedingly profound insight on Carnap's part to realize that inductive logic can, to a large extent anyway, dispense entirely with rules of acceptance and inductive inferences in the ordinary sense. Instead, inductive logic attaches numbers to hypotheses, and these numbers are used to make practical decisions. In some circumstances such numbers, the degrees of confirmation, may serve as fair betting quotients to determine the odds for a fair bet on a given hypothesis. There is no rule that tells one when to accept an hypothesis or when to reject it; instead, there is a rule of practical behavior that prescribes that we so act as to maximize our expec- tation of utility.[2] Hence, inductive logic is simply not concerned with inductive ar- guments (regarded as entities composed of premises and conclusions).

Now, I do not completely agree with Carnap on the issue of acceptance rules in inductive logic; I believe that inductive logic does require some inductive infer- ences.[3] But when it comes to probabilities (weights) of single events, I believe that he is entirely correct. In my view, we must establish by inductive inference prob- ability statements, which I regard as statements about limiting frequencies. But, when we come to apply this probability knowledge to single events, we procure a weight which functions just as Carnap has indicated—as a fair betting quotient or as a value to be used in computing an expectation of utility.[4] Consequently, I maintain, in the context of statistical explanation of individual events, we do not try to establish the explanandum as the conclusion of an inductive argument; instead, we need to es- tablish the weights that would appropriately attach to such explanandum events for purposes of betting and other practical behavior. That is precisely what the partition of the reference class into homogeneous subclasses achieves: it establishes the correct weight to assign to *any* member of A with respect to its being a B. First, one de- termines to which compartment C_k it belongs, and then one adopts the value p_k as the weight. Since we adopted the *multiple homogeneity rule,* we can genuinely han- dle any member of A, not just those which happen to fall into one subclass of the original reference class.

One might ask on what grounds we can claim to have characterized explanation. The answer is this. When an explanation (as herein explicated) has been provided, we know exactly how to regard any A with respect to the property B. We know which ones to bet on, which to bet against, and at what odds. We know precisely what degree of expectation is rational. We know how to face uncertainty about an

A's being a B in the most reasonable, practical, and efficient way. We know every factor that is relevant to an A having property B. We know exactly the weight that should have been attached to the prediction that this A will be a B. We know all of the regularities (universal or statistical) that are relevant to our original question. What more could one ask of an explanation?

There are several general remarks that should be added to the foregoing theory of explanation:

a. It is evident that explanations as herein characterized are nomological. For the frequency interpretation probability statements are statistical generations, and every explanation must contain at least one such generalization. Since an explanation essentially consists of a set of statistical generalizations, I shall call these explanations "statistical" without qualification, meaning thereby to distinguish them from what Hempel has recently called "inductive-statistical."[5] His inductive-statistical explanations contain statistical generalizations, but they are inductive inferences as well.

b. From the standpoint of the present theory, deductive-nomological explanations are just a special case of statistical explanation. If one takes the frequency theory of probability as literally dealing with infinite classes of events, there is a difference between the universal generalization, "All A are B," and the statistical generalization, "$P(A,B) = 1$," for the former admits no As that are not Bs, whereas the latter admits of infinitely many As that are not Bs. For this reason, if the universal generalization holds, the reference class A is homogeneous with respect to B, whereas the statistical generalization may be true even if A is not homogeneous. Once this important difference is noted, it does not seem necessary to offer a special account of deductive-nomological explanations.

c. The problem of symmetry of explanation and prediction, which is one of the most hotly debated issues in discussions of explanation, is easily answered in the present theory. To explain an event is to provide the best possible grounds we could have had for making predictions concerning it. An explanation does not show that the event was to be expected; it shows what sorts of expectations would have been reasonable and under what circumstances it was to be expected. To explain an event is to show to what degree it was to be expected, and this degree may be translated into practical predictive behavior such as wagering on it. In some cases the explanation will show that the explanandum event was not to be expected, but that does not destroy the symmetry of explanation and prediction. The symmetry consists in the fact that the explanatory facts constitute the fullest possible basis for making a prediction of whether or not the event would occur. To explain an event is not to predict it *ex post facto*, but a complete explanation does provide complete grounds for rational prediction concerning that event. Thus, the present account of explanation does sustain a thoroughgoing symmetry thesis, and this symmetry is not refuted by explanations having low weights.

d. In characterizing statistical explanation, I have required that the partition of the reference class yield subclasses that are, in fact, homogeneous. I have not settled for practical or epistemic homogeneity. The question of whether actual homogeneity or epistemic homogeneity is demanded is, for my view, analogous to the question of whether the premises of the explanation must be true or highly confirmed for Hempel's view.[6] I have always felt that truth was the appropriate requirement, for I believe Carnap has shown that the concept of truth is harmless enough.[7] However, for those who feel too uncomfortable with the stricter requirement, it would be possible to characterize statistical explanation in terms of epistemic homogeneity instead of actual homogeniety. No fundamental problem about the nature of explanation seems to be involved.

e. This paper has been concerned with the explanation of single events, but from the standpoint of probability theory, there is no significant distinction between a single event and any finite set of events. Thus, the kind of explanation appropriate to a single result of heads on a single toss of a coin would, in principle, be just like the kind of explanation that would be appropriate to a sequence of ten heads on ten consecutive tosses of a coin or to ten heads on ten different coins tossed simultaneously.

f. With Hempel, I believe that generalizations, both universal and statistical, are capable of being explained. Explanations invoke generalizations as parts of the explanans, but these generalizations themselves may need explanation. This does not mean that the explanation of the particular event that employed the generalization is incomplete; it only means that an additional explanation is possible and may be desirable. In some cases it may be possible to explain a statistical generalization by subsuming it under a higher level generalization; a probability may become an instance for a higher level probability. For example, Reichenbach offered an explanation for equiprobability in games of chance, by constructing, in effect, a sequence of probability sequences.[8] Each of the first level sequences is a single case with respect to the second level sequence. To explain generalizations in this manner is simply to repeat, at a higher level, the pattern of explanation we have been discussing. Whether this is or is not the only method of explaining generalizations is, of course, an entirely different question.

g. In the present account of statistical explanation, Hempel's problem of the "nonconjunctiveness of statistical systematization"[9] simply vanishes. This problem arises because in general, according to the multiplication theorem for probabilities, the probability of a conjunction is smaller than that of either conjunct taken alone. Thus, if we have chosen a value r, such that explanations are acceptable only if they confer upon the explanandum an inductive probability of at least r, it is quite possible that each of the two explananda will satisfy that condition, whereas their conjunction fails to do so. Since the characterization of explanation I am offering makes no demands whatever for high probabilities (weights), it has no problem of nonconjunctiveness.

Conclusion

Although I am hopeful that the foregoing analysis of statistical explanation of single events solely in terms of statistical relevance relations is of some help in understanding the nature of scientific explanation, I should like to cite, quite explicitly, several respects in which it seems to be incomplete.

First, and most obviously, whatever the merits of the present account, no reason has been offered for supposing the type of explanation under consideration to be the only legitimate kind of scientific explanation. If we make the usual distinction between empirical laws and scientific theories, we could say that the kind of explanation I have discussed is explanation by means of empirical laws. For all that has been said in this paper, theoretical explanation—explanation that makes use of scientific theories in the fullest sense of the term—may have a logical structure entirely different from that of statistical explanation. Although theoretical explanation is almost certainly the most important kind of scientific explanation, it does, nevertheless, seem useful to have a clear account of explanation by means of empirical laws, if only as a point of departure for a treatment of theoretical explanation.

Second, in remarking above that statistical explanation is nomological, I was tacitly admitting that the statistical or universal generalizations invoked in explanations should be lawlike. I have made no attempt to analyze lawlikeness, but it seems likely that an adequate analysis will involve a solution to Nelson Goodman's "grue-bleen" problem.[10]

Third, my account of statistical explanation obviously depends heavily upon the concept of *statistical relevance* and upon the *screening-off relation*, which is defined in terms of statistical relevance. In the course of the discussion, I have attempted to show how these tools enable us to capture much of the involvement of explanation with causality, but I have not attempted to provide an analysis of causation in terms of these statistical concepts alone. Reichenbach has attempted such an analysis,[11] but whether his—or any other—can succeed is a difficult question. I should be inclined to harbor serious misgivings about the adequacy of my view of statistical explanation if the statistical analysis of causation cannot be carried through successfully, for the relation between causation and explanation seems extremely intimate.

2. Causal Connections*

Earlier I have made frequent reference to the role of causality in scientific explanation, but I have done nothing to furnish an analysis of the concept of causality or its subsidiary notions. The time has come to focus attention specifically upon this issue, and to see whether we can provide an account of causality adequate for a causal theory of scientific explanation. I shall not attempt to sidestep the fundamental philosophical issues. It seems to me that intellectual integrity demands that we squarely face Hume's incisive critique of causal relations and come to terms with the profound problems he raised.[12]

Basic Problems

As a point of departure for the discussion of causality, it is appropriate for us to take a look at the reasons that have led philosophers to develop theories of explanation that do not require causal components. To Aristotle and Laplace it must have seemed evident that scientific explanations are inevitably causal in character. Laplacian determinism is causal determinism, and I know of no reason to suppose that Laplace made any distinction between causal and noncausal laws. In their 1948 paper, Hempel and Oppenheim make the same sort of identification in an offhand manner (Hempel, 1965, p. 250; but see note 6, same page, added in 1964); however, in subsequent writings, Hempel has explicitly renounced this view (e.g., 1965, pp. 352–54).

It might be initially tempting to suppose that all laws of nature are causal laws, and that explanation in terms of laws is *ipso facto* causal explanation. It is, however, quite easy to find law-statements that do not express causal relations. Many regularities in nature are not direct cause-effect relations. Night follows day, and day follows night; nevertheless, day does not cause night, and night does not cause day. Kepler's laws of planetary motion describe the orbits of the planets, but they offer no causal account of these motions.[13] Similarly, the ideal gas law

$$PV = nRT$$

relates pressure (P), volume (V), and temperature (T) for a given sample of gas, and it tells how these quantities vary as functions of one another, but it says nothing whatever about causal relations among them. An increase in pressure might be brought about by moving a piston so as to decrease the volume, or it might be caused by an increase in temperature. The law itself is entirely noncommittal concerning such causal considerations. Each of these regularities—the alternation of night with day; the regular motions of the planets; and the functional relationship among temperature, pressure, and volume of an ideal gas—can be *explained* causally, but they do not *express* causal relations. Moreover, they do not afford causal explanations of the events subsumed under them. For this reason, it seems to me, their value in providing scientific explanations of particular events is, at best, severely limited. These are regularities that need to be explained, but that do not, by themselves, do much in the way of explaining other phenomena.

To untutored common sense, and to many scientists uncorrupted by philosophical training, it is evident that causality plays a central role in scientific explanation. An appropriate answer to an explanation-seeking why-question normally begins with the word "because," and the causal involvements of the answer are usually not hard to find.[14] The concept of causality has, however, been philosophically suspect ever since David Hume's devastating critique, first published in 1739 in his *Treatise of Human Nature*. In the "Abstract" of that work, Hume wrote:

> Here is a billiard ball lying on the table, and another ball moving toward it with rapidity. They strike; the ball which was formerly at rest now acquires a motion. This is as perfect an instance of the relations of cause and effect as any which we know either by sensation or reflection. Let us therefore examine it. It is evident that the two balls touched one another before the motion was communicated, and that there was no interval betwixt the shock and the motion. *Contiguity* in time and place is therefore a requisite circumstance to the operation of all causes. It is evident, likewise, that the motion which was the cause is prior to the motion which was the effect. *Priority* in time is, therefore, another requisite circumstance in every cause. But this is not all. Let us try any other balls of the same kind in a like situation, and we shall always find that the impulse of the one produces motion in the other. Here, therefore, is a *third* circumstance, viz. that of *constant conjunction* betwixt the cause and the effect. Every object like the cause produces always some object like the effect. Beyond these three circumstances of contiguity, priority, and constant conjunction I can discover nothing in this cause (1955, p. 186–87)

This discussion is, of course, more notable for factors Hume was unable to find than for those he enumerated. In particular, he could not discover any 'necessary connections' relating causes to effects, or any 'hidden powers' by which the cause "brings about" the effect. This classic account of causation is rightly regarded as a landmark in philosophy.

In an oft-quoted remark that stands at the beginning of a famous 1913 essay, Bertrand Russell warns philosophers about the appeal to causality:

> All philosophers, of every school, imagine that causation is one of the fundamental axioms or postulates of science, yet, oddly enough, in advanced sciences such as gravitational astronomy, the word "cause" never occurs. . . . To me it seems that . . . the reason why physics has ceased to look for causes is that, in fact, there are no such things. The law of causality, I believe, like much that passes muster among philosophers, is a relic of a bygone age, surviving, like the monarchy, only because it is erroneously supposed to do no harm (1929, p. 180)

It is hardly surprising that, in the light of Hume's critique and Russell's resounding condemnation, philosophers with an empiricist bent have been rather wary of the use of causal concepts. By 1927, however, when he wrote *The Analysis of Matter*, Russell recognized that causality plays a fundamental role in physics; in *Human Knowledge*, four of the five postulates he advanced as a basis for all scientific knowledge make explicit reference to causal relations (1948, pp. 487–96). It should be noted, however, that the causal concepts he invokes are *not* the same as the traditional philosophical ones he had rejected earlier.[15] In contemporary physics, causality is a pervasive ingredient (Suppes, 1970, pp. 5–6).

Two Basic Concepts

A standard picture of causality has been around at least since the time of Hume. The general idea is that we have two (or more) distinct events that bear some sort of cause-effect relations to one another. There has, of course, been considerable controversy regarding the nature of both the relation and the relata. It has sometimes been maintained, for instance, that facts or propositions (rather than events) are the sorts of entities that can constitute relata. It has long been disputed whether causal relations can be said to obtain among individual events, or whether statements about cause-effect relations implicitly involve assertions about classes of events. The relation itself has sometimes been taken to be that of sufficient condition, sometimes necessary condition, or perhaps a combination of the two.[16] Some authors have even proposed that certain sorts of statistical relations constitute causal relations.

The foregoing characterization obviously fits J. L. Mackie's sophisticated account in terms of INUS conditions—that is, *insufficient but non-redundant* parts of *unnecessary* but *sufficient* conditions (1974, p. 62). The idea is this. There are several different causes that might account for the burning down of a house: careless smoking in bed, an electrical short circuit, arson, being struck by lightning. With certain obvious qualifications, each of these may be taken as a sufficient condition for the fire, but none of them can be considered necessary. Moreover, each of the sufficient conditions cited involves a fairly complex combination of conditions, each of which constitutes a nonredundant part of the particular sufficient condition under consideration. The careless smoker, for example, must fall asleep with his cigarette, and it must fall upon something flammable. It must not awaken the smoker by burning him before it falls from his hand. When the smoker does become aware of the fire, it must have progressed beyond the stage at which he can extinguish it. Any one of these necessary components of some complex sufficient condition can, under certain circumstances, qualify as a cause. According to this standard approach, events enjoy the status of fundamental entities, and these entities are "connected" to one another by cause-effect relations.

It is my conviction that this standard view, in all of its well-known variations, is profoundly mistaken, and that a radically different notion should be developed. I shall not, at this juncture, attempt to mount arguments against the standard conception [. . .]. Instead, I shall present a rather different approach for purposes of comparison. I hope that the alternative will stand on its own merits.

There are, I believe, two fundamental causal concepts that need to be explicated, and if that can be achieved, we will be in a position to deal with the problems of causality in general. The two basic concepts are *propagation* and *production*, and both are familiar to common sense. The first of these will be treated in this [part]; the second will be handled in the next [part]. When we say that the blow of a hammer drives a nail, we mean that the impact produces penetration of the nail into the wood. When we say that a horse pulls a cart, we mean that the force exerted by the horse produces the motion of the cart. When we say that lightning ignites a forest, we mean that the electrical discharge produces a fire. When we say that a person's embarrassment was due to a thoughtless remark, we mean that an inappropriate comment produced psychological discomfort. Such examples of causal production occur frequently in everyday contexts.

Causal propagation (or transmission) is equally familiar. Experiences that we had earlier in our lives affect our current behavior. By means of memory, the influence of these past events is transmitted to the present (see Rosen, 1975). A sonic boom makes us aware of the passage of a jet airplane overhead; a disturbance in the air is propagated from the upper atmosphere to our location on the ground. Signals transmitted from a broadcasting station are received by the radio in our home. News or music reaches us because electromagnetic waves are propagated from the transmitter to the receiver. In 1775, some Massachusetts farmers—in initiating the American Revolutionary War—"fired the shot heard 'round the world" (Emerson, 1836). As all of these examples show, what happens at one place and time can have significant influence upon what happens at other places and times. This is possible because causal influence can be propagated through time and space. Although causal production and causal propagation are intimately related to one another, we should, I believe, resist any temptation to try to reduce one to the other.

Processes

One of the fundamental changes that I propose in approaching causality is to take processes rather than events as basic entities. I shall not attempt any rigorous definition of processes; rather I shall cite examples and make some very informal remarks. The main difference between events and processes is that events are relatively localized in space and time, while processes have much greater temporal duration, and in many cases, much greater spatial extent. In space-time diagrams, events are represented by points, while processes are represented by lines. A baseball colliding with a window would count as an event; the baseball, traveling from the bat to the window, would constitute a process. The activation of a photocell by a pulse of light would be an event; the pulse of light, traveling, perhaps from a distant star, would be a process. A sneeze is an event. The shadow of a cloud moving across the landscape is a process. Although I shall deny that all processes qualify as causal processes, what I mean by a process is similar to what Russell characterized as a causal line:

> A causal line may always be regarded as the persistence of something—a person, a table, a photon, or what not. Throughout a given causal line, there may be constancy of quality, constancy of structure, or a gradual change of either, but not sudden changes of any considerable magnitude. (1948, p. 459)

Among the physically important processes are waves and material objects that persist through time. As I shall use these terms, even a material object at rest will qualify as a process.

Before attempting to develop a theory of causality in which processes, rather than events, are taken as fundamental, I should consider briefly the scientific legitimacy of this approach. In Newtonian mechanics, both spatial extent and temporal duration were absolute quantities. The length of a rigid rod did not depend upon a choice of frame of reference, nor did the duration of a process (such as the length of time between the creation and destruction of a material object). Given two events, in Newtonian mechanics, both the spatial distance and the temporal separation between them were absolute magnitudes. A 'physical thing ontology' was thus appropriate to classical physics. As everyone knows, Einstein's special theory of relativity changed all that. Both the spatial distance and the temporal separation were relativized to frames of reference. The length of a rigid rod and the duration of a temporal process varied from one frame of reference to another. However, as Minkowski showed, there is an invariant quantity—the space-time interval between two events. This quantity is independent of the frame of reference; for any two events, it has the same value in each and every inertial frame of reference. Since there are good reasons for according a fundamental physical status to invariants, it was a natural consequence of the special theory of relativity to regard the world as a collection of events that bear space-time relations to one another. These considerations offer support for what is sometimes called an 'event ontology.'

There is, however, another way (originally developed by A. A. Robb) of approaching the special theory of relativity; it is done entirely with paths of light pulses. At any point in space-time, we can construct the Minkowski light cone—a two-sheeted cone whose surface is generated by the paths of all possible light pulses that converge upon the point (past light cone) and the paths of all possible light pulses that could be emitted from the point (future light cone). When all of the light cones are given, the entire space-time structure of the world is determined (see Winnie, 1977). But light pulses, traveling through space and time, are processes. We can, therefore, base special relativity upon a 'process ontology.' Moreover, this approach can be extended in a natural way to general relativity by taking into account the paths of freely falling material particles; these moving gravitational test particles are also processes (see Grünbaum, 1973, pp. 735–50). It is, consequently, entirely legitimate to approach the space-time structure of the physical world by regarding physical processes as the basic types of physical entities. The theory of relativity does not mandate an 'event ontology.'

Whether one adopts the event-based approach or the process-based approach, causal relations must be accorded a fundamental place in the special theory of relativity. As we have seen, any given event E_0, occurring at a particular space-time point P_0, has an associated double-sheeted light cone. All events that could have a causal influence upon E_0 are located in the interior or on the surface of the past light cone, and all events upon which E_0 could have any causal influence are located in the interior or on the surface of the future light cone. All such events are *causally connectable* with E_0. Those events that lie on the surface of either sheet of the light cone are said to have a *lightlike separation* from E_0, those that lie within either part of the cone are said to have a *timelike separation* from E_0, and those that are outside of the cone are said to have a *spacelike separation* from E_0. The Minkowski light cone can, with complete propriety, be called "the cone of causal relevance," and the entire space-time structure of special relativity can be developed on the basis of causal concepts (Winnie, 1977).

Special relativity demands that we make a distinction between *causal processes* and *pseudo-processes*. It is a fundamental principle of that theory that light is a *first signal*—that is, no signal can be transmitted at a velocity greater than the velocity of light in a vacuum. There are, however, certain processes that can transpire at arbitrarily high velocities—at velocities vastly exceeding that of light. This fact does not violate the basic relativistic principle, however, for these 'processes' are incapable of serving as signals or of transmitting information. Causal processes are those that are capable of transmitting signals; pseudo-processes are incapable of doing so.

Consider a simple example. Suppose that we have a very large circular building—a sort of super-Astrodome, if you will—with a spotlight mounted at its center. When the light is turned on in the otherwise darkened building, it casts a spot of light upon the wall. If we turn the light on for a brief moment, and then off again, a light pulse travels from the light to the wall. This pulse of light, traveling from the spotlight to the wall, is a paradigm of what we mean by a causal process. Suppose, further, that the spotlight is mounted on a mechanism that makes it rotate. If the light is turned on and set into rotation, the spot of light that it casts upon the wall will move around the outer wall in a highly regular fashion. This 'process'— the moving spot of light—seems to fulfill the conditions Russell used to characterize causal lines, but it is not a causal process. It is a paradigm of what we mean by a pseudo-process.

The basic method for distinguishing causal processes from pseudo-processes is the criterion of mark transmission. A causal process is capable of transmitting a mark; a pseudo-process is not. Consider, first, a pulse of light that travels from the spotlight to the wall. If we place a piece of red glass in its path at any point between the spotlight and the wall, the light pulse, which was white, becomes and remains red until it reaches the wall. A single intervention at one point in the process transforms it in a way that persists from that point on. If we had not intervened, the light pulse would have remained white during its entire journey from the spotlight to the wall. If we do intervene locally at a single place, we can produce a change that is transmitted from the point of intervention onward. We shall say, therefore, that the light pulse constitutes a causal process whether it is modified or not, since in either case it is capable of transmitting a mark. Clearly, light pulses can serve as signals and can transmit messages; remember Paul Revere, "One if by land and two if by sea."

Now, let us consider the spot of light that moves around the wall as the spotlight rotates. There are a number of ways in which we can intervene to change the spot at some point; for example, we can place a red filter at the wall with the result that the spot of light becomes red at that point. But if we make such a modification in the traveling spot, it will not be transmitted beyond the point of interaction. As soon as the light spot moves beyond the point at which the red filter was placed, it will become white again. The mark can be made, but it will not be transmitted. We have a 'process,' which, in the absence of any intervention, consists of a white spot moving regularly along the wall of the building. If we intervene at some point, the 'process' will be modified *at that point*, but it will continue on beyond that point just as if no intervention had occurred. We can, of course, make the spot red at other places if we wish. We can install a red lens in the spotlight, but that does not constitute a *local* intervention at an isolated point in the process itself. We can put red filters at many places along the wall, but that would involve *many* interventions rather than a single one. We could get someone to run around the wall holding a red filter in front of the spot continuously, but that would not constitute an intervention *at a single point* in the 'process.'

This last suggestion brings us back to the subject of velocity. If the spot of light is moving rapidly, no runner could keep up with it, but perhaps a mechanical device could be set up. If, however, the spot moves too rapidly, it would be physically impossible to make the filter travel fast enough to keep pace. No material object, such as the filter, can travel at a velocity greater than that of light, but no such limitation is placed upon the spot on the wall. This can easily be seen as follows. If the spotlight rotates at a fixed rate, then it takes the spot of light a fixed amount of time to make one entire circuit around the wall. If the spotlight rotates once per second, the spot of light will travel around the wall in one second. This fact is independent of the size of the building. We can imagine that without making any change in the spotlight or its rate of rotation, the outer walls are expanded indefinitely. At a certain point, when the radius of the building is a little less than 50,000 kilometers, the spot will be traveling at the speed of light (300,000 km/sec). As the walls are moved still farther out, the velocity of the spot exceeds the speed of light.

To make this point more vivid, consider an actual example that is quite analogous to the rotating spotlight. There is a pulsar in the crab nebula that is about 6,500 light-years away. This pulsar is thought to be a rapidly rotating neutron star that sends out a beam of radiation. When the beam is directed toward us, it sends out radiation that we detect later as a pulse. The pulses arrive at the rate of 30 per second; that is the rate at which the neutron star rotates. Now, imagine a circle drawn with the pulsar at its center, and with a radius equal to the distance from the pulsar to the earth. The electromagnetic radiation from the pulsar (which travels at the speed of light) takes 6,500 years to traverse the radius of this circle, but the "spot" of radiation sweeps around the circumference of this circle in 1/30th of a second; at that rate, it is traveling at about 4×10^{13} times the speed of light. There is no upper limit on the speed of pseudo-processes.[17]

Another example may help to clarify this distinction. Consider a car traveling along a road on a sunny day. As the car moves at 100 km/hr, its shadow moves along the shoulder at the same speed. The moving car, like any material object, constitutes a causal process; the shadow is a pseudo-process. If the car collides with a stone wall, it will carry the marks of that collision—the dents and scratches—

along with it long after the collision has taken place. If, however, only the shadow of the car collides with the stone wall, it will be deformed momentarily, but it will resume its normal shape just as soon as it has passed beyond the wall. Indeed, if the car passes a tall building that cuts it off from the sunlight, the shadow will be obliterated, but it will pop right back into existence as soon as the car has returned to the direct sunlight. If, however, the car is totally obliterated—say, by an atomic bomb blast—it will not pop back into existence as soon as the blast has subsided.

A given process, whether it be causal or pseudo, has a certain degree of uniformity—we may say, somewhat loosely, that it exhibits a certain structure. The difference between a causal process and a pseudo-process, I am suggesting, is that the causal process transmits its own structure, while the pseudo-process does not. The distinction between processes that do and those that do not transmit their own structures is revealed by the mark criterion. If a process—a causal process—is transmitting its own structure, then it will be capable of transmitting certain modifications in that structure.

In *Human Knowledge*, Russell placed great emphasis upon what he called "causal lines," which he characterized in the following terms:

> A "causal line," as I wish to define the term, is a temporal series of events so related that, given some of them, something can be inferred about the others whatever may be happening elsewhere. A causal line may always be regarded as the persistence of something—a person, table, a photon, or what not. Throughout a given causal line, there may be constancy of quality, constancy of structure, or gradual change in either, but not sudden change of any considerable magnitude. (1948, p. 59)

He then goes on to comment upon the significance of causal lines:

> That there are such more or less self-determined causal processes is in no degree logically necessary, but is, I think, one of the fundamental postulates of science. It is in virtue of the truth of this postulate—if it is true—that we are able to acquire partial knowledge in spite of our enormous ignorance. (Ibid.)

Although Russell seems clearly to intend his causal lines to be what we have called causal processes, his characterization may appear to allow pseudo-processes to qualify as well. Pseudo-processes, such as the spot of light traveling around the wall of our Astrodome, sometimes exhibit great uniformity, and their regular behavior can serve as a basis for inferring the nature of certain parts of the pseudo-process on the basis of observation of other parts. But pseudo-processes are not self-determined; the spot of light is determined by the behavior of the beacon and the beam it sends out. Moreover, the inference from one part of the pseudo-process to another is *not* reliable *regardless of what may be happening elsewhere,* for if the spotlight is switched off or covered with an opaque hood, the inference will go wrong. We may say, therefore, that our observations of the various phenomena going on in the world around us reveal processes that exhibit considerable regularity, but some of these are genuine causal processes and others are pseudo-processes. The causal processes are, as Russell says, self-determined; they transmit their own uniformities of qualitative and structural features. The regularities exhibited by the pseudo-processes, in contrast, are parasitic upon causal regularities exterior to the 'process' itself—in the case of the Astrodome, the behavior of the beacon; in the case of the shadow traveling along the roadside, the behavior of the car and the sun. The ability to transmit a mark is the criterion that distinguishes causal processes from pseudo-processes, for if the modification represented by the mark is propagated, the process is transmitting its own characteristics. Otherwise, the 'process' is not self-determined, and is not independent of what goes on elsewhere.

Although Russell's characterization of causal lines is heuristically useful, it cannot serve as a fundamental criterion for their identification for two reasons. First, it is formulated in terms of our ability to infer the nature of some portions from a knowledge of other portions. We need a criterion that does not rest upon such epistemic notions as knowledge and inference, for the existence of the vast majority of causal processes in the history of the universe is quite independent of human knowers. This aspect of the characterization could, perhaps, be restated nonanthropocentrically in terms of the persistence of objective regularities in the process. The second reason is more serious. To suggest that processes have regularities that persist "whatever may be happening elsewhere" is surely an overstatement. If an extremely massive object should happen to be located in the neighborhood of a light pulse, its path will be significantly altered. If a nuclear blast should occur in the vicinity of a mail truck, the letters that it carries will be totally destroyed. If sunspot activity reaches a high level, radio communication is affected. Notice that, in each of these cases, the factor cited does not occur or exist on the world line of the process in question. In each instance, of course, the disrupting factor initiates processes that intersect with the process in question, but that does not undermine the objection to the claim that causal processes transpire in their self-determined fashion regardless of what is happening elsewhere. A more acceptable statement might be that a causal process would persist even if it were isolated from external causal influences. This formulation, unfortunately, seems at the very least to flirt with circularity, for external causal influences must be transmitted to the locus of the process in question by means of other processes. We shall certainly want to keep clearly in mind the notion that causal processes are not parasitic upon other processes, but it does not seem likely that this rough idea could be transformed into a useful basic criterion.

It has often been suggested that the principal characteristic of causal processes is that they transmit energy. While I believe it is true that all and only causal processes transmit energy, there is, I think, a fundamental problem involved in employing this fact as a basic criterion—namely, we must have some basis for distinguishing situations in which energy is transmitted from those in which it merely appears in some regular fashion. The difficulty is easily seen in the "Astrodome" example. As a light pulse travels from the central spotlight to the wall, it carries radiant energy; this energy is present in the various stages of the process as the pulse travels from the lamp to the wall. As the spot of light travels around the wall, energy appears at the places occupied by the spot, but we do not want to say that this energy is transmitted. The problem is to distinguish the cases in which a given bundle of energy is transmitted through a process from those in which different bundles of energy are appearing in some regular fashion. The key to this distinction is, I believe, the mark method. Just as the detective makes his mark on the murder weapon for purposes of later identification, so also do we make marks in processes so that the energy present at one space-time locale can be identified when it appears at other times and places.

A causal process is one that is self-determined and not parasitic upon other causal influences. A causal process is one that transmits energy, as well as information and causal influence. The fundamental criterion for distinguishing self-determined energy transmitting processes from pseudo-processes is the capability of such processes of transmitting marks. In the next section, we shall deal with the concept of transmission in greater detail.

Our main concern with causal processes is their role in the propagation of causal influences; radio broadcasting presents a clear example. The transmitting station sends a carrier wave that has a certain structure—characterized by amplitude and frequency, among other things—and modifications of this wave, in the form of modulations of amplitude (AM) or frequency (FM), are imposed for the purpose of broadcasting. Processes that transmit their own structures are capable of transmitting marks, signals, information, energy, and causal influence. Such processes are the means by which causal influence is propagated in our world. Causal influences, transmitted by radio, may set your foot to tapping, or induce someone to purchase a different brand of soap, or point a television camera aboard a spacecraft toward the rings of Saturn. A causal influence transmitted by a flying arrow can pierce an apple on the head of William Tell's son. A causal influence transmitted by sound waves can make your dog come running. A causal influence transmitted by ink marks on a piece of paper can gladden one's day or break someone's heart.

It is evident, I think, that the propagation or transmission of causal influence from one place and time to another must play a fundamental role in the causal structure of the world. As I shall argue next, causal processes constitute precisely the causal connections that Hume sought, but was unable to find.

The 'At-At' Theory of Causal Propagation

In the preceding section, I invoked Reichenbach's mark criterion to make the crucial distinction between causal processes and pseudo-processes. Causal processes are distinguished from pseudo-processes in terms of their ability to transmit marks. In order to qualify as causal, a process need not actually be transmitting a mark; the requirement is that it be capable of doing so.

When we characterize causal processes partly in terms of their ability to transmit marks, we must deal explicitly with the question of whether we have violated the kinds of strictures Hume so emphatically expounded. He warned against the uncritical use of such concepts as 'power' and 'necessary connection.' Is not the *ability to transmit* a mark an example of just such a mysterious power? Kenneth Sayre expressed his misgivings on this score when, after acknowledging the distinction between causal interactions and causal processes, he wrote:

> The causal process, continuous though it may be, is made up of individual events related to others in a causal nexus. . . . it is by virtue of the relations among the members of causal series that we are enabled to make the inferences by which causal processes are characterized. . . . if we do not have an adequate conception of the relatedness between individual members in a causal series, there is a sense in which our conception of the causal process itself remains deficient. (1977, p. 206)

The 'at-at' theory of causal transmission is an attempt to remedy this deficiency.

Does this remedy illicitly invoke the sort of concept Hume proscribed? I think not. Ability to transmit a mark can be viewed as a particularly important species of constant conjunction—the sort of thing Hume recognized as observable and admissible. It is a matter of performing certain kinds of experiments. If we place a red filter in a light beam near its source, we can observe that the mark—redness—appears at all places to which the beam is subsequently propagated. This fact can be verified by experiments as often as we wish to perform them. If, contrariwise (returning to our Astrodome example of the preceding section), we make the spot on the wall red by placing a filter in the beam at one point just before the light strikes the wall (or by any other means we may devise), we will see that the mark—redness—is not present at all other places in which the moving spot subsequently appears on the wall. This, too, can be verified by repeated experimentation. Such facts are straightforwardly observable.

The question can still be reformulated. What do we mean when we speak of *transmission*? How does the process *make* the mark appear elsewhere within it? There is, I believe, an astonishingly simple answer. The transmission of a mark from point *A* in a causal process to point *B* in the same process *is* the fact that it appears at each point between *A* and *B* *without further interactions*. If *A* is the point at which the red filter is inserted into the beam going from the spotlight to the wall, and *B* is the point at which the beam strikes the wall, then only the interaction at *A* is required. If we place a white card in the beam at any point between *A* and *B*, we will find the beam red at that point.

The basic thesis about mark transmission can now be stated (in a principle I shall designate MT for "mark transmission") as follows:

> MT: Let *P* be a process that, in the absence of interactions with other processes, would remain uniform with respect to a characteristic *Q*, which it would manifest consistently over an interval that includes both of the space-time points *A* and *B* (*A* ≠ *B*). Then, a mark (consisting of a modification of *Q* into *Q'*), which has been introduced into process *P* by means of a single local interaction at point *A*, is transmitted to point *B* if *P* manifests the modification *Q'* at *B* and at all stages of the process between *A* and *B* without additional interventions.

This principle is clearly counterfactual, for it states explicitly that the process *P* would have continued to manifest the characteristic *Q* if the specific marking interaction had not occurred. This subjunctive formulation is required, I believe, to overcome an objection posed by Nancy Cartwright (in conversation) to previous formulations. The problem is this. Suppose our rotating beacon is casting a white spot that moves around the wall, and that we mark the spot by interposing a red filter at the wall. Suppose further, however, that a red lens has been installed in the beacon just a tiny fraction of a second earlier, so that the spot on the wall becomes red at the moment we mark it with our red filter, but it remains red from that point on because of the red lens. Under these circumstances, were it not for the counterfactual condition, it would appear that we had satisfied the requirement formulated in MT, for we have marked the spot by a single interaction at point *A*, and the spot remains red from that point on to any other point *B* we care to designate, without any additional interactions. As we have just mentioned, the installation of the red lens on the spotlight does not constitute a marking of the spot on the wall. The counterfactual stipulation given in the first sentence of MT <u>blocks</u> situations of the sort mentioned by Cartwright, in which we would most certainly want to deny that any mark transmission occurred via the spot moving around the wall. In this case, the moving spot would have turned red because of the lens even if no marking interaction had occurred locally at the wall.

A serious misgiving arises from the use of counterfactual formulations to characterize the distinction between causal processes and pseudo-processes; it concerns the question of objectivity. The distinction is fully objective. It is a matter of fact that a light pulse constitutes a causal process, while a shadow is a pseudo-process. Philosophers have often maintained, however, that counterfactual conditionals involve unavoidably pragmatic aspects. Consider the famous example about Verdi and Bizet. One person might say, "If Verdi had been a compatriot of Bizet, then Verdi would have been French," whereas another might maintain, "If Bizet had been a compatriot of Verdi, then Bizet would have been Italian." These two statements seem incompatible with one another. Their antecedents are logically equivalent; if, however, we accept both conditionals, we wind up with the conclusion that Verdi would be French, that Bizet would be Italian, and they would still not be compatriots. Yet both statements can be true. The first person could be making an unstated presupposition that the nationality of Bizet is fixed in this context, while the second presupposes that the nationality of Verdi is fixed. What remains fixed and what is subject to change—which are established by pragmatic features of the context in which the counterfactual is uttered—determine whether a counterfactual is true or false. It is concluded that counterfactual conditional statements do not express objective facts of nature; indeed, van Fraassen (1980, p. 118) goes so far as to assert that science contains no counterfactuals. If that sweeping claim were true (which I seriously doubt),[18] the foregoing criterion MT would be in serious trouble.

Although MT involves an explicit counterfactual, I do not believe that the foregoing difficulty is insurmountable. Science has a direct way of dealing with the kinds of counterfactual assertions we require, namely, the experimental approach. In a well-designed controlled experiment, the experimenter determines which conditions are to be fixed for purposes of the experiment and which allowed to vary. The result of the experiment establishes some counterfactual statements as true and others as false under well-specified conditions. Consider the kinds of cases that concern us; such counterfactuals can readily be tested experimentally. Suppose we want to see whether the beam traveling from the spotlight to the wall is capable of transmitting the red mark. We set up the following experiment. The light will be turned on and off one hundred times. At a point midway between the spotlight and the wall, we station an experimenter with a random number generator. Without communicating with the experimenter who turns the light on and off, this second experimenter uses his device to make a random selection of fifty trials in which he will make a mark and fifty in which he will not. If all and only the fifty instances in which the marking interaction occurs are those in which the spot on the wall is red, as well as all the intervening stages in the process, then we may conclude with reasonable certainty that the fifty cases in which the beam was red subsequent to the marking interaction are cases in which the beam would not have been red if the marking interaction had not occurred. On any satisfactory analysis of counterfactuals, it seems to me, we would be justified in drawing such a conclusion. It should be carefully noted that I am *not* offering the foregoing experimental procedure as an analysis of counterfactuals; it is, indeed, a result that we should expect any analysis to yield.

A similar experimental approach could obviously be taken with respect to the spot traversing the wall. We design an experiment in which the beacon will rotate one hundred times, and each traversal will be taken as a separate process. We station an experimenter with a random number generator at the wall. Without communicating with the experimenter operating the beacon, the one at the wall makes a random selection of fifty trials in which to make the mark and fifty in which to refrain. If it turns out that some or all of the trials in which no interaction occurs, are, nevertheless, cases in which the spot on the wall turns red as it passes the second experimenter, then we know that we are *not* dealing with cases in which the process will not turn from white to red if no interaction occurs. Hence, if in some cases the spot turns red and remains red after the mark is imposed, we know we are not entitled to conclude that the mark has actually been transmitted.

The account of mark transmission embodied in principle MT—which is the proposed foundation for the concept of propagation of causal influence—may seem too trivial to be taken seriously. I believe such a judgment would be mistaken. My reason lies in the close parallel that can be drawn between the foregoing solution to the problem of mark transmission and the solution of an ancient philosophical puzzle.

About twenty-five hundred years ago, Zeno of Elea enunciated some famous paradoxes of motion, including the well-known paradox of the flying arrow. This paradox was not adequately resolved until the early part of the twentieth century. To establish an intimate connection between this problem and our problem of causal transmission, two observations are in order. First, a physical object (such as the arrow) moving from one place to another constitutes a causal process, as can be demonstrated easily by application of the mark method—for example, initials carved on the shaft of the arrow before it is shot are present on the shaft after it hits its target. And there can be no doubt that the arrow propagates causal influence. The hunter kills his prey by releasing the appropriately aimed arrow; the flying arrow constitutes the causal connection between the cause (release of the arrow from the bow under tension) and the effect (death of a deer). Second, Zeno's paradoxes were designed to prove the absurdity not only of motion, but also of every kind of process or change. Henri Bergson expressed this point eloquently in his discussion of what he called "the cinematographic view of becoming." He invites us to consider any process, such as the motion of a regiment of soldiers passing in review. We can take many snapshots—static views—of different stages of the process, but, he argues, we cannot really capture the movement in this way, for,

> every attempt to reconstitute change out of states implies the absurd proposition, that movement is made out of immobilities.
>
> Philosophy perceived this as soon as it opened its eyes. The arguments of Zeno of Elea, although formulated with a very different intention, have no other meaning.
>
> Take the flying arrow. (1911, p. 308; quoted in Salmon, 1970a, p. 63).

Let us have a look at this paradox. At any given instant, Zeno seems to have argued, the arrow is where it is, occupying a portion of space equal to itself. During the instant it cannot move, for that would require the instant to have parts, and an instant is *by definition* a minimal and indivisible element of time. If the arrow did move during the instant, it would have to be in one place at one part of the instant and in a different place at another part of the instant. Moreover, for the arrow to move during the instant would require that during that instant it must occupy a space larger than itself, for otherwise it has no room to move. As Russell said:

> It is never moving, but in some miraculous way the change of position has to occur *between* the instants, that is to say, not at any time whatever. This is what M. Bergson calls the cinematographic representation of reality. The more the difficulty is meditated, the more real it becomes (1929, p. 187; quoted in Salmon, 1970a, p. 51)

There is a strong temptation to respond to this paradox by pointing out that the differential calculus provides us with a perfectly meaningful definition of instantaneous velocity, and that this quantity *can* assume values other than zero. Velocity is change of position with respect to time, and the derivative dx/dt furnished an expression that can be evaluated for particular values of t. Thus an arrow can be at rest at a given moment—that is, dx/dt may equal 0 for that particular value of t. Or it can be in motion at a given moment—that is, dx/dt might be 100 km/hr for another particular value of t. Once we recognize this elementary result of the infinitesimal calculus, it is often suggested, the paradox of the flying arrow vanishes.

This appealing attempt to resolve the paradox is, however, unsatisfactory, as Russell clearly realized. The problem lies in the definition of the derivative; dx/dt is defined as the limit as Δt approaches 0 of $\Delta x/\Delta t$, where Δt represents a non-zero interval of time and Δx may be a non-zero spatial distance. In other words, instantaneous velocity is defined as the limit, as we take decreasing time intervals, of the noninstantaneous average velocity with which the object traverses what is—in the case of non-zero values— a non-zero stretch of space. Thus in the definition of instantaneous velocity, we employ the concept of non-instantaneous velocity, which is precisely the problematic concept from which the paradox arises. To put the same point in a different way, the concept of instantaneous velocity does not genuinely characterize the motion of an object at an isolated instant all by itself, for the very definition of instantaneous velocity makes reference to neighboring instants of time and neighboring points of space. To find an adequate resolution of the flying arrow paradox, we must go deeper.

To describe the motion of a body, we express the relation between its position and the moments of time with which we are concerned by means of a mathematical function; for example, the equation of motion of a freely falling particle near the surface of the earth is

$$(1) \qquad\qquad x = f(t) = 1/2gt^2$$

where $g = 9.8$ m/sec^2. We can therefore say that this equation furnishes a function $f(t)$ that relates the position of x to the time t. But what is a mathematical function? It is a set of pairs of numbers; for each admissible value of t, there is an associated value of x. To say that an object moves in accordance with equation (1) is simply to say that *at* any given moment t it is *at* point x, where the correspondence between the values of t and of x is given by the set of pairs of numbers that constitute the function represented by equation (1). To move from point *A* to point *B* is simply to be *at* the appropriate point of space *at* the appropriate moment of time—no more, no less. The resulting theory is therefore known as "the 'at-at' theory of motion." To the best of my knowledge, it was first clearly formulated and applied to the arrow paradox by Russell.

According to the 'at-at' theory, to move from *A* to *B* is simply to occupy the intervening points at the intervening instants. It consists in being *at* particular points of space *at* corresponding moments. There is no *additional* question as to how the arrow *gets from* point *A* to point *B;* the answer has already been given—by being at the intervening points at the intervening moments. The answer is emphatically *not* that it gets from *A* to *B* by zipping through the intermediate points at high speed. Moreover, there is no additional question about how the arrow gets from one intervening point to another—the answer is the same, namely, by being at the points between them at the corresponding moments. And clearly, there can be no question about how the arrow gets from one point to the next, for in a continuum there is no next point. I am convinced that Zeno's arrow paradox is a profound problem concerning the nature of change and motion, and that its resolution by Russell in terms of the 'at-at' theory of motion represents a distinctly nontrivial achievement.[19] The fact that this solution can—if I am right—be extended in a direct fashion to provide a resolution of the problem of mark transmission is an additional laurel.

The 'at-at' theory of mark transmission provides, I believe, an acceptable basis for the mark method, which can in turn serve as the means to distinguish causal processes from pseudo-processes. The world contains a great many types of causal processes—transmission of light waves, motion of material objects, transmissions of sound waves, persistence of crystalline structure, and so forth. Processes of any of these types may occur without having any mark imposed. In such instances, the processes still qualify as causal. *Ability* to transmit a mark is the criterion of causal processes; processes that are *actually* unmarked may be causal. Unmarked processes exhibit some sort of persistent structure, as Russell pointed out in his characterization of causal lines; in such cases, we say that the structure is transmitted within the causal process. Pseudo-processes may also exhibit persistent structure; in these cases, we maintain that the structure is *not transmitted* by means of the "process" itself, but by some other external agency.

The basis for saying that the regularity in the causal process is transmitted via the process itself lies in the ability of the causal process to transmit a modification in its structure—a mark—resulting from an interaction. Consider a brief pulse of white light; it consists of a collection of photons of various frequencies, and if it is not polarized, the waves will have various spatial orientations. If we place a red filter in the path of this pulse, it will absorb all photons with frequencies falling outside of the red range, allowing only those within that range to pass. The resulting pulse has its structure modified in a rather precisely specifiable way, and the fact that this modification persists is precisely what we mean by claiming that the mark is transmitted. The counterfactual clause in our principle MT is designed to rule out structural changes brought about by anything other than the marking interaction. The light pulse could, alternatively, have been passed through a polarizer. The resulting pulse would consist of photons having a specified spatial orientation instead of the miscellaneous assortment of orientations it contained before encountering the polarizer. The principle of structure transmission (ST) may be formulated as follows:

ST: If a process is capable of transmitting changes in structure due to marking interactions, then that process can be said to transmit its own structure.

The fact that a process does not transmit a particular type of mark, however, does not mean that it is not a causal process. A ball of putty constitutes a causal process, and one kind of mark it will transmit is a change in shape imposed by indenting it with the thumb. However, a hard rubber ball is equally a causal process, but it will not transmit the same sort of mark, because of its elastic properties. The fact that a particular sort of structural modification does not persist, because of some inherent tendency of the process to resume its earlier structure, does not mean it is not transmitting its own structure; it means only that we have not found the appropriate sort of mark for that kind of process. A hard rubber ball can be marked by painting a spot on it, and that mark will persist for a while.

Marking methods are sometimes used in practice for the identification of causal processes. As fans of Perry Mason are aware, Lieutenant Tragg always placed "his mark" upon the murder weapon found at the scene of the crime in order to be able to identify it later at the trial of the suspect. Radioactive tracers are used in the investigation of physiological processes—for example, to determine the course taken by a particular substance ingested by a subject. Malodorous substances are added to natural gas used for heating and cooking in order to ascertain the presence of leaks; in fact, one large chemical manufacturer published full-page color advertisements in scientific magazines for its product "La Stink."

One of the main reasons for devoting our attention to causal processes is to show how they can transmit causal influence. In the case of causal processes used to transmit signals, the point is obvious. Paul Revere was caused to start out on his famous night ride by a light signal sent from the tower of the Old North Church. A drug, placed surreptitiously in a drink, can cause a person to lose consciousness because it retains its chemical structure as it is ingested, absorbed, and circulated through the body of the victim. A loud sound can produce a painful sensation in the ears because the disturbance of the air is transmitted from the origin to the hearer. Radio signals sent to orbiting satellites can activate devices aboard because the wave retains its form as it travels from earth through space. The principle of propagation of causal influence (PCI) may be formulated as follows:

PCI: A process that transmits its own structure is capable of propagating a causal influence from one space-time locale to another.

The propagation of causal influence by means of causal processes *constitutes*, I believe, the mysterious connection between cause and effect which Hume sought.

In offering the 'at-at' theory of mark transmission as a basis for distinguishing causal processes from pseudo-processes, we have furnished an account of the transmission of information and propagation of causal influence without appealing to any of the secret powers which Hume's account of causation soundly proscribed. With this account we see that the mysterious connection between causes and effects is not very mysterious after all.

Our task is by no means finished, however, for this account of transmission of marks and propagation of causal influence has used the unanalyzed notion of a causal interaction that produces a mark. Unless a satisfactory account of causal interaction and mark production can be provided, our theory of causality will contain a severe lacuna. We will attempt to fill that gap in the next [section]. Nevertheless, we have made significant progress in explicating the fundamental concept, introduced at the beginning of the chapter, of *causal propagation* (or *transmission*).

This [section] is entitled "Causal Connections," but little has actually been said about the way in which causal processes provide the connection between cause and effect. Nevertheless, in many common-sense situations, we talk about causal relations between pairs of spatiotemporally separated events. We might say, for instance, that turning the key causes the car to start. In this context we assume, of course, that the electrical circuitry is intact, that the various parts are in good working order, that there is gasoline in the tank, and so forth, but I think we can make sense of a cause-effect relation only if we can provide a *causal connection* between the cause and the effect. This involves tracing out the causal processes that lead from the turning of the key and the closing of an electrical circuit to various occurrences that eventuate in the turning over of the engine and the ignition of fuel in the cylinders. We say, for another example, that a tap on the knee causes the foot to jerk. Again, we believe that there are neural impulses traveling from the place at which the tap occurred to the muscles that control the movement of the foot, and processes in those muscles that lead to movement of the foot itself. The genetic relationship between parents and offspring provides a further example. In this case, the molecular biologist refers to the actual process of information via the DNA molecule employing the genetic code.

In each of these situations, we analyze the cause-effect relations in terms of three components—an event that constitutes the cause, another event that constitutes the effect, and a causal process that connects the two events. In some cases, such as the starting of the car, there are many intermediate events, but in such cases, the successive intermediate events are connected to one another by spatiotemporally continuous causal processes. A splendid example of multiple causal connections was provided by David Kaplan. Several years ago, he paid a visit to Tucson, just after completing a boat trip through the Grand Canyon with his family. The best time to take such a trip, he remarked, is when it is very hot in Phoenix. What is the causal connection to the weather in Phoenix, which is about 200 miles away? At such times, the air conditioners in Phoenix are used more heavily, which places a greater load on the generators at the Glen Canyon Dam (above the Grand Canyon). Under the circumstances, more water is allowed to pass through the turbines to meet the increased demand for power, which produces a greater flow of water down the Colorado River. This results in a more exciting ride through the rapids in the Canyon.

In the next [section], we shall consider events—especially causal interactions—more explicitly. It will then be easier to see how causal processes constitute precisely the physical connections between causes and effects that Hume sought—what he called "the cement of the universe." These causal connections will play a vital role in our account of scientific explanation.

It is tempting, of course, to try to reduce causal processes to chains of events; indeed, people frequently speak of causal chains. Such talk can be seriously misleading if it is taken to mean that causal processes are composed of discrete events that are serially ordered so that any given event has an immediate successor. If, however, the continuous character of causal processes is kept clearly in mind, I would not argue that it is philosophically incorrect to regard processes as collections of events. At the same time, it does seem heuristically disadvantageous to do so, for this practice seems almost inevitably to lead to the puzzle (articulated by Sayre in the quotation given previously) of how these events, which make up a given process, are causally related to one another. The point of the 'at-at' theory, it seems to me is to show that no such question about the causal relations among the constituents of the process need arise—for the same reason that, aside from occupying intermediate positions at the appropriate times, there is no further question about how the flying arrow gets from one place to another. With the aid of the 'at-at' theory, we have a complete answer to Hume's penetrating question about the nature of causal connections. For this heuristic reason, then, I consider it advisable to resist the temptation always to return to formulations in terms of events.

3. Causal Forks and Common Causes

There is a familiar pattern of causal reasoning that we all use every day, usually without being consciously aware of it. Confronted with what appears to be an improbable coincidence, we seek a common cause. If the common cause can be found, it is invoked to explain the coincidence.

Conjunctive Forks

Suppose, for example, that several members of a traveling theatrical company who have spent a pleasant day in the country together become violently ill that evening. We infer that it was probably due to a common meal of which they all partook. When we find that their lunch included some poisonous mushrooms that they had gathered and cooked, we have the explanation. There is a certain small chance that a particular actor or actress will, on any given evening, suffer severe gastrointestinal distress—the probability need not, of course, be the same for each person. If the illnesses were statistically independent of one another, then the probability of all the picnickers becoming ill on the same night would be equal to the product of all of these individual small probabilities. Even though a chance coincidence of this sort is possible, it is too improbable to be accepted as such. The evidence for a common cause is compelling.

Although reasoning of this type seems simple and straightforward, philosophers have paid surprisingly little explicit attention to it. Hans Reichenbach is the outstanding exception. In his posthumous book, *The Direction of Time* (1956), he enunciated *the principle of the common cause,* and he attempted to explicate the principle in terms of a statistical structure that he called a *conjunctive fork.* The principle of the common cause states, roughly, that when apparent coincidences occur that are too improbable to be attributed to chance, they can be explained by reference to a common causal antecedent. This principle is by no means trivial or vacuous. Among other things, it denies that such apparent coincidences are to be explained teleologically in terms of subsequent common effects. If the aforementioned theatrical troupe had been scheduled to put on a performance that evening, it would in all likelihood have been canceled. This common effect would not, however, explain the coincidence. We shall have to consider why this is so later [. . .].

Other examples, from everyday life and from science, are easy to find. If, for instance, two students in a class turn in identical papers, and if we can rule out the

Suppose no mushrooms, belief that the audience will be horrid

possibility that either copied directly from the other, then we search for a common cause—for example, a paper in a fraternity file from which both of them copied independently of each other.

A recent astronomical discovery, which has considerable scientific significance, furnished a particularly fine example. The twin quasars 0975 + 561 A and B are separated by an angular width of 5.7 sec of arc. Two quasars in such apparent proximity would be a rather improbable occurrence given simply the observed distribution of quasars. Examination of their spectra indicates equal red shifts, and hence, equal distances. Thus these objects are close together in space, as well as appearing close together as seen from earth. Moreover, close examination of their spectra reveals a striking similarity—indeed, they are indistinguishable. This situation is in sharp contrast to the relations between the spectra of any two quasars picked at random. Astronomers immediately recognized the need to explain this astonishing coincidence in terms of some sort of common cause. One hypothesis that was entertained quite early was that twin quasars had somehow (no one had the slightest idea how this could happen in reality) developed from a common ancestor. Another hypothesis was the gravitational lens effect—that is, there are not in fact two distinct quasars, but the two images were produced from a single body by the gravitational bending of the light by an intervening massive object. This result might be produced by a black hole, it was theorized, or by a very large elliptical galaxy. Further observation, under fortuitously excellent viewing conditions, has subsequently revealed the presence of a galaxy that would be adequate to produce the gravitational splitting of the image. This explanation is now, to the best of my knowledge, accepted by virtually all of the experts (Chaffee, 1980).

In an attempt to characterize the structure of such examples of common causes, Reichenbach (1956, section 19) introduced the notion of a *conjunctive fork*, defined in terms of the following four conditions.[20]

(1) $$P(A.B|C) = P(A|C) \times P(B|C)$$

(2) $$P(A.B|\bar{C}) = P(A|\bar{C}) \times P(B|\bar{C})$$

(3) $$P(A|C) > P(A|\bar{C})$$

(4) $$P(B|C) > P(B|\bar{C}).$$

For reasons that will be made clear [. . .], we shall stipulate that none of the probabilities occurring in these relations is equal to zero or one. Although it is not immediately obvious, conditions (1)–(4) entail

(5) $$P(A.B) > P(A) \times P(B)$$

(see Reichenbach, 1956, pp. 160–61).[21] These relations apply quite straightforwardly in concrete situations. Given two effects, A and B, that occur together more frequently than they would if they were statistically independent of one another, there is some prior event C, which is a cause of A and is also a cause of B, that explains the lack of independence between A and B. In the case of plagiarism, the cause C is the presence of the term paper in the file to which both students had access. In the case of simultaneous illness, the cause C is the common meal that included the poisonous mushrooms. In the case of the twin quasar image, the cause C is the emission of radiation in two slightly different directions by a single luminous body.

To say of two events, X and Y, that they occurred independently of one another means that they occur together with a probability equal to the product of the probabilities of their separate occurrences; that is,

(6) $$P(X.Y) = P(X) \times P(Y).$$

Thus in the examples we have considered, as relation (5) states, the two effects A and B are not independent. However, given the occurrence of the common cause C, A, and B do occur independently, as the relationship among the conditional probabilities in equation (1) shows. Thus in the case of illness, the fact that the probability of two individuals being ill at the same time is greater than the product of the probabilities of their individual illnesses is explained by the common meal. In this example, we are assuming that the fact that one person is afflicted does not have any direct causal influence upon the illness of the other.[22] Moreover, let us assume for the sake of simplicity that in this situation, there are no other potential common causes of severe gastrointestinal illness.[23] Then, in the absence of the common cause C—that is, when $\bar{C}$ obtains—A and B are also independent of one another, as the relationship among the conditional probabilities in equation (2) states. Relations (3) and (4) simply assert that C is a positive cause of A and B, since the probability of each is greater in the presence of C than in the absence of C.

There is another useful way to look at equations (1) and (2). Recalling that, according to the multiplication theorem,

(7) $$P(A.B|C) = P(A|C) \times P(B|A.C),$$

we see that, provided $P(A|C) \neq 0$, equation (1) entails

(8) $$P(B|C) = P(B|A.C).$$

In Reichenbach's terminology, this says that C screens off A from B. A similar argument shows that $\bar{C}$ screens off B from A. To screen off *means* to make statistically irrelevant. Thus, according to equation (1), the common cause C makes each of the two effects A and B statistically irrelevant to one another. By applying the same argument to equation (2), we can easily see that it entails that the absence of the common cause also screens off A from B.

To make quite clear the nature of the conjunctive fork, I should like to use an example deliberately contrived to exhibit the relationships involved. Suppose we have a pair of dice that are rolled together. If the first die comes to rest with side 6 on the top, that is an event of the type A; if the second die comes to rest with side 6 uppermost, that is an event of type B. These dice are like standard dice except for the fact that each one has a tiny magnet embedded in it. In addition, the table on which they are thrown has a powerful electromagnet beneath its surface. This magnet can be turned on or off with a concealed switch. If the dice are rolled when the electromagnet is on, it is considered an instance of the common cause C; if the magnet is off when the dice are tossed, the event is designated as $\bar{C}$. Let us further assume that when the electromagnet is turned off, these dice behave exactly as standard dice. The probability of getting 6 with either dies is 1/6, and the probability of getting double 6 is 1/36.[24] If the electromagnet is turned on, let us assume, the chance of getting 6 with either die is 1/2, and the probability of double 6 is 1/4. It is easily seen that conditions (1)–(4) are fulfilled. Let us make a further stipulation, which will simplify the arithmetic, but which has no other bearing upon the essential features of the example—namely, that half of the tosses of this pair of dice are made with the electromagnet turned on, and half are made with it turned off. We might imagine some sort of random device that controls the switch, and that realizes this equiprobability condition. We can readily see that the overall probability of 6 on each die, regardless of whether the electromagnet is on or off, is 1/3. In addition, the overall probability of double 6 is the arithmetical average of 1/4 and 1/36, which equals 5/36. If the occurrence of 6 on one die were independent of 6 occurring on the other, the overall probability of double 6 would be $1/3 \times 1/3 = 1/9 \neq 5/36$. Thus the example satisfies relation (5), as of course it must, in addition to relations (1)–(4).

It may initially seem counterintuitive to say that the results on the two dice are statistically independent if the electromagnet is off, and they are statistically independent if it is on, but that overall they are not independent. But they are, indeed, nonindependent, and this nonindependence arises from a clustering of 6's, which is due simply to the fact that in a subset of the class of all tosses the probability of 6 is enhanced for both dice. Thus dependency arises, not because of any physical interaction between the dice, but because of special background conditions that obtain on certain of the tosses but not on others. The same consideration applies to the earlier, less contrived, cases. When the two students each copy from a paper in a fraternity file, there is no direct physical interaction between the process by which one of the papers is produced and that by which the other is produced—in fact, if either student had been aware that the other was using that source, the unhappy coincidence might have been avoided. Likewise, as explicitly mentioned in the mushroom poisoning case—where, to make the example fit formulas (1)–(4), we confine attention to just two of the performers—the illness of one of them had no effect upon the illness of the other. The coincidence resulted from the fact that a common set of background conditions obtained, namely, a common food supply from which both ate. Similarly, in the twin quasar example, the two images are formed by two separate radiation processes that come from a common source, but do not directly interact with each other anywhere along the line.

Reichenbach claimed—correctly, I believe—that conjunctive forks possess an important asymmetry. Just as we can have two effects that arise out of a given common cause, so also may we find a common effect resulting from two distinct causes. For example, by getting results on a roll of two dice that add up to 7, one may win a prize. Reichenbach distinguished three situations: (1) a common cause C giving rise to two separate effects, A and B, without any common effect arising from A and B conjointly; (2) two events A and B that, in the absence of a common cause C, jointly produce a common effect E; and (3) a combination of (1) and (2) in which the events A and B have both a common cause C and a common effect E. He characterized situations (1) and (2) as *open forks,* while (3) is closed on both ends. Reichenbach's asymmetry thesis was that situations of type (2) never represent conjunctive forks; conjunctive forks that are open are always open to the future and never to the past. Since the statistical relations found in conjunctive forks are said

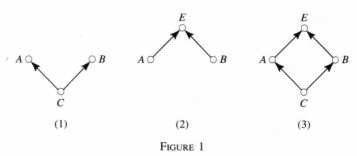

FIGURE 1

to explain otherwise improbable coincidences, it follows that such coincidences are explained only in terms of common causes, never common effects. In the case of a prize being awarded for the result 7 on a toss of two dice, we do not explain the occurrence of 7 in terms of the awarding of the prize. This is not a mere philosophical prejudice against teleological explanations. Assuming a fair game, we believe, in fact, that the probability of getting a 7 is the same regardless of whether a prize is involved. In situations of type (2), in which there is no common cause to produce a statistical dependency, A and B occur independently; the common effect E, unlike a common cause, does not create a correlation between A and B. This is a straightforward factual assertion, which can, in cases like the tossing of dice, be tested empirically. If, contrary to expectation, we should find that the result 7 does not occur with a probability of $1/6$ in cases where a prize is at stake, we could be confident that the (positive or negative) correlation between the outcomes on the two dice was a result of some prior tampering—recall the magnetic dice, where the electromagnet had to be turned on before the dice came to rest to affect the probability of the result—rather than "events conspiring" to reward one player or to prevent another from receiving a benefit. A world in which teleological causation operates is not logically impossible, but our world does not seem, as a matter of fact, to be of such a kind.

In order to appreciate fully the import of Reichenbach's asymmetry thesis, let us look at an initially plausible putative counterexample provided by Frank Jackson.[25] It will be instructive to make an explicit comparison between his example and a bona fide instance of a common cause. Let us begin with the common cause. Suppose that two siblings contract mumps at the same time, and assume that neither caught the disease from the other. The coincidence is explained by the fact that they attended a birthday party and, by virtue of being in the same locale, both were exposed to another child who had the disease. This would constitute a typical example of a conjunctive fork.

Now, with that kind of example in mind, consider a case that involves Hansen's disease (leprosy). One of the traditional ways of dealing with this illness was by segregating its victims in colonies. Suppose that Adams has Hansen's disease (A) and Baker also has it (B). Previous to contracting the disease, Adams and Baker had never lived in proximity to one another, and there is no victim of the disease with whom both had been in contact. We may therefore assume that there is no common cause. Subsequently, however, Adams and Baker are transported to a colony, where both are treated with chaulmoogra oil (the traditional treatment). The fact that both Adams and Baker are in the colony and exposed to chaulmoogra oil is a common effect of the fact that each of them has Hansen's disease. This situation, according to Jackson, constitutes a conjunctive fork A, E, B, where we have a common effect E, but no common cause. We must see whether it does, in fact, qualify. It is easy to see that relations (3) and (4) are satisfied, for

$$P(A|E) > P(A|\bar{E})$$

and

$$P(B|E) > P(B|\bar{E}),$$

that is, the probability of Adams having Hansen's disease is greater if he and Baker are members of the colony than it would be if they were not. If not both Adams and Baker are members of the colony, it might be that Adams is a member and Baker is not (in which case the probability that Adams has the disease is high), but it might also be that Baker is a member and Adams is not, or that neither of them is (in which cases the probability that Adams has the disease would be very low). The same reasoning holds for Baker, *mutatis mutandis*.

The crucial question concerns relations (1) and (2). Substituting "E" for "C" in those two equations, let us recall that they say, respectively, that A and B are statistically independent of one another, given that condition E holds, and they are statistically indpendent when E does not hold. Now, if A and B are independent of one another, it follows immediately that their negations $\bar{A}$ and $\bar{B}$ are independent; thus relation (1) implies

$$P(\bar{A}.\bar{B}|E) = P(\bar{A}|E) \times P(\bar{B}|E).$$

This tells us, as we saw previously, that E screens off $\bar{A}$ from $\bar{B}$, that is,

$$P(\bar{B}|E) = P(\bar{B}|\bar{A}.E).$$

Let us therefore ask whether the fact that both Adams and Baker are members of the colony (and both are exposed to chaulmoogra oil) would make the fact that Adams did not have Hansen's disease statistically irrelevant to Baker's failure to have that disease. The answer is clearly negative. Among the members of the colony, a small percentage—doctors, nurses, missionaries—do not have Hansen's disease, and those involved in actual treatment of the victims are exposed to chaulmoogra oil. Suppose, for example, that Adams and Baker both belong to the colony and are exposed to chaulmoogra oil, but that Baker does not have leprosy. To make the situation concrete, suppose that there are one hundred members of the colony who are exposed to chaulmoogra oil, and among them are only two medical personnel who do not have Hansen's disease. If Baker has Hansen's disease, the probability that Adams does not have it is about 0.02, while if Baker does not have it, the probability for Adams is about 0.01—a difference of a factor of two. As stipulated in this example, the fact that Adams has the disease has no direct causal relevance to the fact that Baker also has it, and conversely, but given the circumstances specified, they are statistically relevant to one another.

Although we know that A, B, and E do not form a conjunctive fork, let us look also at relation (2). For this purpose, we shall ask whether $\bar{E}$ screens off A and B from each other—that is, whether

$$P(B|\bar{E}) = P(B|A.\bar{E}).$$

Let us suppose, for the sake of this argument, that there is only one colony for victims of Hansen's disease, and that almost every afflicted person belongs to it—indeed, let us assume for the moment that only two people in the world have Hansen's disease who have not yet joined the colony. Thus, if a pair of people are chosen at random, and if not both belong to the colony, it is quite unlikely that even one of them has Hansen's disease. However, among the many pairs not both of whom are in the colony, but at least one of whom has Hansen's disease, there is only one consisting of two individuals who both have this disease. Thus, under our extreme temporary assumption about the colony, it is clear that $\bar{E}$ does not screen off A from B. Given that not both Adams and Baker are members of the colony receiving the traditional treatment, it is quite unlikely that Baker has Hansen's disease, but given in addition that Adams has the disease, it becomes even more unlikely that Baker has it. The assumption that there are only two victims of Hansen's disease in the world who are not members of the colony is, of course, altogether unrealistic; however, given the relative rarity of the disease among the population at large, the failure of screening off, though not as dramatic, must still obtain. We see from the failure of relations (1) and (2) that Jackson's example does not constitute a conjunctive fork. In spite of the great superficial similarity between the mumps and leprosy situations, there is a deep difference between them. This comparison exemplifies Reichenbach's asymmetry thesis.

Although Reichenbach held only that there are no conjunctive forks that are open toward the past, I believe that an even stronger claim is warranted—though I shall merely illustrate it but not try to argue it here. I am inclined to believe that conjunctive forks, whether open or closed by a fourth event, always point in the same temporal direction. Reichenbach allowed that in situations of type (3), the two events A and B along with their common effect E could form a conjunctive fork. Here, of course, there must also be a common cause C, and it is C rather than E that explains the coincidental occurrence of A and B. I doubt that, even in these circumstances, A, B, and E can form a conjunctive fork.

Suppose—to return to the mushroom poisoning example mentioned previously—that among the afflicted troupers are the leading lady and leading man, and that the performance scheduled for that evening is canceled. I shall assume, as before, that the two illnesses along with the common cause form a conjunctive fork. The analysis that shows that the fork consisting of the two illnesses and the cancellation of the performance is not conjunctive is similar to the analysis of the leprosy example. Let us focus upon equation (2), and consider it from the standpoint of screening off. We must compare the probability that the leading lady is ill, given that the performance is not canceled, with the probability that the leading lady is ill, given that the leading man is ill and the performance is not canceled. It is implausible to suppose that the two are equal, for it seems much more likely that the show could go on if only one of the leading characters is ill than it would be if both were ill. It should be recalled that we are discussing a traveling theatrical company, so we should not presume that a large complement of stand-ins is available. Therefore, the performance of the play does not screen off the one illness from the other, and relation (2) in the definition of the conjunctive fork does not hold. One could use a similar argument to show, in this case, that relation (1) is also violated.

I do not have any compelling argument to prove that in the case of the double fork [in the previous figure], A, C, B and A, E, B cannot both be conjunctive. Such combinations are logically possible, for one can find sets of probability values that satisfy the eight relations—four for each fork.[26] Nevertheless, I have not been able to find or construct a physically plausible example, and Reichenbach never furnished one. It would be valuable to know whether double conjunctive forks of this sort would violate some basic physical principle. I do not know the answer to this question.

Reichenbach's principle of the common cause asserts the existence of common causes in cases of improbable coincidences, but it does not assert that *any* event C that fulfills relations (1)–(4) qualifies *ipso facto* as a common cause of A and B.[27] The following example, due to Ellis Crasnow, illustrates this point. Consider a man who usually arrives at his office about 9:00 A.M., makes a cup of coffee, and settles down to read the morning paper. On some occasions, however, he arrives promptly at 8:00 A.M., and on these very same mornings his secretary has arrived somewhat earlier and prepared a fresh pot of coffee. Moreover, on just these mornings, he is met at his office by one of his associates who normally works at a different location. Now, if we consider the fact that the coffee is already made when he arrives (A) and the fact that his associate shows up on that morning (B) as the coincidence to be explained, then it might be noted that on such mornings he always catches the 7:00 A.M. bus (C), while on other mornings he usually takes the 8:00 A.M. bus ($\bar{C}$). In this example, it is plausible enough to suppose that A, B, and C form a conjunctive fork satisfying (1)–(4), but obviously C cannot be considered a cause either of A or of B. The actual common cause is an entirely different event C', namely a telephone appointment made the day before by his secretary. C' is, in fact, the common cause of A, B, and C.

This example leaves us with an important question. Given an event C that, along with A and B, forms a conjunctive fork, how can we tell whether C is a bona fide common cause? The answer, I think, is quite straightforward. C must be connected to A and B by suitable causal processes of the sort discussed in the preceding [part]. These causal processes constitute the mechanisms by which causal influence is transmitted from the cause to each of the effects.

Interactive Forks

There is another, basically different, sort of common cause situation that cannot appropriately be characterized in terms of conjunctive forks. Consider a simple example. Two pool balls, the cue ball and the 8-ball, lie upon a pool table. A relative novice attempts a shot that is intended to put the 8-ball into one of the far corner pockets, but given the positions of the balls, if the 8-ball falls into one corner pocket, the cue ball is almost certain to go into the other far corner pocket, resulting in a "scratch." Let A stand for the 8-ball dropping into the one corner pocket, let B stand for the cue ball dropping into the other corner pocket, and let C stand for the collision between the cue ball and the 8-ball that occurs when the player executes the shot. We may reasonably assume that the probability of the 8-ball going into the pocket is $1/2$ if the player tries the shot, and that the probability of the cue ball going into the pocket is also about $1/2$. It is immediately evident that A, B, and C do not constitute a conjunctive fork, for C does not screen off A and B from one another. Given that the shot is attempted, the probability that the cue ball will fall into the pocket (approximately $1/2$) is not equal to the probability that the cue ball will go into the pocket, given that the shot has been attempted and that the 8-ball has dropped into the other far corner pocket (approximately 1).

In discussing the conjunctive fork, I took some pains to point out that forks of that sort occur in situations in which separate and distinct processes, which do not directly interact, arise out of special background conditions. In the example of the pool balls, however, there is a direct interaction—a collision—between the two causal processes consisting of portions of the histories of the two balls. For this reason, I have suggested that forks that are exemplified by such cases be called *interactive forks*.[28] Since the common cause C does not statistically screen off the two effects A and B from one another, interactive forks violate condition (1) in the definition of conjunctive forks.

The best way to look at interactive forks, I believe, is in terms of spatiotemporal intersections of processes. In some cases, two processes may intersect without producing any lasting modification in either. This will happen, for example, when both processes are pseudo-processes. If the paths of two airplanes, flying in different directions at different altitudes on a clear day, cross one another, the shadows on the ground may coincide momentarily. But as soon as the shadows have passed the intersection, both move on as if no such intersection had ever occurred. In the case of the two pool balls, however, the intersection of their paths results in a change in the motion of each that would not have occurred if they had not collided. Energy and momentum are transferred from one to the other; their respective states of motion are altered. Such modifications occur, I shall maintain, only when (at least) two causal processes intersect. If either or both of the intersecting processes are pseudo-processes, no such mutual modification occurs. However, it is entirely possible for two causal processes to intersect without any subsequent modification in either. Barring the extremely improbable occurrence of a particle-particle type collision between two photons, light rays normally pass right through one another without any lasting effect upon either one of them. The fact that two intersecting processes are both causal is a necessary but not sufficient condition of the production of lasting changes in them.

When two causal processes intersect and suffer lasting modifications after the intersection, there is some correlation between the changes that occur in them. In many cases—and perhaps all—energy and/or momentum transfer occurs, and the correlations between the modifications are direct consequences of the respective conservation laws.[29] This is illustrated by the Compton scattering of an energetic photon off of an electron that can be considered, for practical purposes, initially at rest. The difference in energy between the incoming photon $h\nu$ and the scattered photon $h\nu'$ is equal to the kinetic energy of the recoiling electron. Similarly, the momentum change in the photon is exactly compensated by the momentum change in the electron.[30]

When two processes intersect, and they undergo correlated modifications that persist after the intersection, I shall say that the intersection constitutes a *causal interaction*. This is the basic idea behind what I want to take as a fundamental causal concept. Let C stand for the event consisting of the intersection of two processes. Let A stand for a modification in one and B for a modification in the other. Then, in many cases, we find a relation analogous to equation (1) in the definition of the conjunctive fork, except that the equality is replaced by an inequality:

$$(9) \qquad P(A.B|C) > P(A|C) \times P(B|C).$$

Moreover, given a causal interaction of the foregoing sort, I shall say that the change in each process is *produced* by the interaction with the other process.

I have now characterized, at least partially, the two fundamental causal concepts mentioned at the beginning of [section 2]. Causal processes are the means by which causal influence is *propagated,* and changes in processes are *produced* by causal interactions. We are now in a position to see the close relationship between these basic notions. The distinction between causal processes and pseudo-processes was formulated in terms of the criterion of mark transmission. A mark is a modification in a process, and if that mark persists, the mark is transmitted. Modifications in processes occur when they intersect with other processes; if the modifications persist beyond the point of intersection, then the intersection constitutes a causal interaction and the interaction has produced marks that are transmitted. For example, a pulse of white light is a process, and a piece of red glass is another process. If these two processes intersect—that is, if the light pulse goes through the red glass—then the light pulse becomes and remains red, while the filter undergoes an increase in energy as a result of absorbing some of the light that impinges upon it. Although the newly acquired energy may soon be dissipated into the surrounding environment, the glass retains some of the added energy for some time beyond the actual moment of interaction.

We may, therefore, turn the presentation around in the following way. We live in a world which is full of processes (causal or pseudo), and these processes undergo frequent intersections with one another. Some of these intersections constitute causal interactions; others do not. Let us attempt to formulate a principle CI (for causal interaction) that will set forth the condition explicitly:

CI: Let P_1 and P_2 be two processes that intersect with one another at the space-time point S, which belongs to the histories of both. Let Q be a characteristic that process P_1 would exhibit throughout an interval (which includes subintervals on both sides of S in the history of P_1) if the intersection with P_2 did not occur; let R be a characteristic that process P_2 would exhibit throughout an interval (which includes subintervals on both sides of S in the history of P_2) if the intersection with P_2 did not occur. Then, the intersection of P_1 and P_2 at S constitutes a causal interaction if:

(1) P_1 exhibits the characteristic Q before S, but it exhibits a modified characteristic Q' throughout an interval immediately following S; and

(2) P_2 exhibits the characteristic R before S, but it exhibits a modified characteristic R' throughout an interval immediately following S.

The modifications that Q and R undergo will normally—perhaps invariably—be correlated to one another in accordance with some conservation law, but it seems unnecessary to include this as a requirement in the definition.

This principle, like the principle MT (mark transmission), is formulated in counterfactual terms, for reasons similar to those that induced us to employ a counterfactual formulation in MT. Consider the following example. Two spots of light, one red and the other green, are projected on a white screen. The red spot moves diagonally across the screen from the lower left-hand corner to the upper right-hand corner, and the green spot moves diagonally from the lower right-hand corner to the upper left-hand corner. Let these spots be projected in such a way that they meet and merge momentarily at the center of the screen; at that moment, a yellow spot appears (the result of combining red and green light—mixing colored light is altogether different from mixing colored paints), but each resumes its former color as soon as it leaves the region of intersection. Since no modification of color persists beyond the intersection, we are not tempted to suppose, on the basis of this observation, that a causal interaction has occurred.

Now let us modify the set up. Again, we have the two spots of light projected upon the screen, but in the new situation they travel in different paths. The red spot moves diagonally from the lower left-hand corner to the center of the screen, and then it travels from the center to the upper left-hand corner. The green spot moves from the lower right-hand corner to the center, and then to the upper right-hand corner. Assuming, as before, that the two spots of light meet at the center of the screen, we could describe what we see in either of two ways. First, we could say that the red spot collides with the green spot in the center of the screen, and the directions of motion of the two spots are different after the collision. Second, we could say that the spot that travels from the lower left to the upper right changes from red to green as it goes through the intersection in the middle of the screen, while the spot that travels from lower right to upper left changes from green to red as it goes through the intersection. It seems that each spot changes color in the intersection, and the change persists beyond the locus of the intersection. Under either of these descriptions, it may appear that we have observed a causal interaction, but such an appearance is illusory. Two pseudo-processes intersected, but no causal interaction occurred.

The counterfactual formulation of principle CI is designed to deal with examples of this sort. Under the conditions specified in the physical setup, the red spot moves from the lower left corner to the center and then to the upper left corner of the screen regardless of whether the green spot is present or not. It would be false to say that the red spot would have traveled across the screen from lower left to upper right if it had not met the green spot. Parallel remarks apply, *mutatis mutandis,* to the behavior of the green spot. Similarly, it would be false to say that the color of the spot that traveled from lower left to upper right would not have changed color if it had not encountered the spot traveling from lower right to upper left. CI is not vulnerable to putative counterexamples of this sort.

Examples of another sort, which were presented independently by Patrick Maher and Richard Otte, are more difficult to handle. Suppose (to use Otte's version) that two billiard balls roll across a table with a transparent surface and that they collide with one another in the center, with the result that their directions of motion are changed. This is, of course, a bona fide causal interaction, and it qualifies as such under CI. We are entitled to say that the direction of motion of the one ball would not have changed in the middle of the table if the collision with the second had not occurred. It is easy to see how that counterfactual could be tested in a controlled experiment. Assume, further, that because of a bright light above the table the billiard balls cast shadows on the floor. When the balls collide the shadows meet, and their directions of motion are different after the intersection. In this case, we would appear to be entitled to say that the direction of motion of the one shadow would not have changed if it had not encountered the other shadow. It looks as if the intersection of the two shadows qualifies as a causal interaction according to CI.

In order to handle examples of this kind, we must consider carefully how our counterfactuals are to be interpreted. This issue arose in connection with the principle of mark transmission (MT) in [part 2], and we made appeal in that context to the testing of counterfactual assertions by means of controlled experiments. We must, I think, approach the question in the same way in this context. We must therefore ask what kind of controlled experiment would be appropriate to test the assertion about the shadows of the colliding billiard balls.

If we just sit around watching the shadows, we will notice some cases in which a shadow, encountering no other shadows, moves straight along the floor. We will notice other cases in which two shadows intersect, and their directions of motions are different after the interesection than they were before. So we can see a correlation between alterations of direction of motion and intersections with other shadows. This procedure, however, consists merely of the collection of available data; it hardly constitutes experimentation. So we must give further thought to the design of some experiments to test our counterfactuals.

Let us formulate the assertion that is to be tested as follows: If shadow #1 had not met shadow #2, then shadow #1 would have continued to move in a straight line instead of changing directions. In order to perform a controlled experiment, we need a situation in which the phenomena under investigation occur repeatedly. Let us mark off a region of the floor that is to be designated as the experimental region. Assume that we have many cases in which two shadows enter that region moving in such a way that they will meet within it, and that their directions of motion after the intersection are different from their prior directions of motion. How this is to be accomplished is no mystery. We could study the behavior of the shadows that occur as many games of billiards are played, or we could simply arrange for an experimenter to roll balls across the table in such a way that they collide with one another in the desired place. Consider one hundred such occurrences. Using some random device, we select fifty of them to be members of the experimental group and fifty to be members of the control group. That selection is not communicated to the experimenter who is manipulating the balls on the table top. When an event in the control group occurs, we simply do nothing and observe the outcome. When an event in the experimental group occurs, we choose one of the entering shadows—call it shadow #2—and shine a light along the path it would have taken if the light had not been directed toward it. This extra illumination obliterates shadow #2; nevertheless, shadow #1 changes its direction in the experimental cases just as it does in the control cases. We have thereby established the falsity of the counterfactual that was to be tested. It is not true that the direction of travel of shadow #1 would not have changed if it had not encountered shadow #2. We have fifty instances in which shadow #1 changed its direction in the absence of shadow #2. The intersection of the two shadows thus fails, according to CI, to qualify as a causal interaction.

As formulated, CI states only a sufficient condition for a causal interaction, since there might be other characteristics, F and G, that suffer the requisite mutual modification even if Q and R do not. In order to transform CI into a condition that is necessary as well as sufficient, we need simply to say that a causal interaction occurs if and only if there exist characteristics Q and R that fulfill the conditions stated previously. It should be noted that the statistical relation (9), which may be a true statement about a given causal interaction, does not enter into the definition of causal interactions.[31]

If two processes intersect in a manner that qualifies as a casual interaction, we may conclude that both processes are causal, for each has been marked (that is, modified) in the intersection with the other and each process transmits the mark beyond the point of intersection. Thus each fulfills the criterion MT; each process shows itself capable of transmitting marks since each one has transmitted a mark generated in the intersection. Indeed, the operation of marking a process is accomplished by means of a causal interaction with another process. Although we may often take an active role in producing a mark in order to ascertain whether a process is causal (or for some other purpose), it should be obvious that human agency plays no essential part in the characterization of causal processes or causal interactions. We have every reason to believe that the world abounded in causal processes and causal interactions long before there were any human agents to perform experiments.

Relations Between Conjunctive and Interactive Forks

Suppose that we have a shooting gallery with a number of targets. The famous sharpshooter Annie Oakley comes to this gallery, but it presents no challenge to her, for she can invariably hit the bull's-eye of any target at which she aims. So, to make the situation interesting, a hardened steel knife-edge is installed in such a position that a direct hit on the knife-edge will sever the bullet in a way that makes one fragment hit the bull's-eye of target A while the other fragment hits the bull's-eye of target B. If we let A stand for a fragment striking the bull's-eye of target A, B for a fragment striking the bull's-eye of target B, and C for the severing of the bullet by the knife-edge, we have an interactive fork quite analogous to the example of the pool balls. Indeed, we may use the same probability values, setting $P(A|C) = P(B|C) = 1/2$, while $P(A|C.B) = P(B|C.A) \simeq 1$. Statistical screening off obviously fails.

We might, however, consider another event C^*. To make the situation concrete, imagine that we have installed between the knife-edge and the targets a steel plate with two holes in it. If the shot at the knife edge is good, then the two fragments of the bullet will go through the two holes, and each fragment will strike its respective bull's-eye with probability virtually equal to 1. Let C^* be the event of the two fragments going through their respective holes. Then, we may say, A, B, and C^* will form a conjunctive fork. That happens because C^* refers to a situation that is subsequent to the physical interaction between the parts of the bullet. By the time we get to C^*, the bullet has been cut into two separate pieces, and each is going its way independently of the other. Even if we should decide to vaporize one of the fragments with a powerful laser, that would have no effect upon the probability of the other fragment finding its target. This example makes quite vivid, I believe, the distinction between the interactive fork, which characterizes direct physical interactions, and the conjunctive fork, which characterizes independent processes arising under special background conditions.[32]

There is a further important point of contrast between conjunctive and interactive forks. Conjunctive forks possess a kind of temporal asymmetry, which was described previously. Interactive forks do not exhibit the same sort of temporal asymmetry. This is easily seen by considering a simple collision between two billiard balls. A collision of this type can occur in reverse; if a collision C precedes states of motion A and B in the two balls, then a collision C can occur in which states of motion just like A and B, except that the direction of motion is reversed, precede the collision. Causal interactions and causal processes do not, in and of themselves, provide a basis for temporal asymmetry.

Our ordinary causal language is infused with temporal asymmetry, but we should be careful in applying it to basic causal concepts. If, for example, we say that two processes are modified as a result of their interaction, the words suggest that we have already determined which are the states of the processes prior to the interaction, and which are the subsequent states. To avoid begging temporal questions, we should say that two processes intersect, and each of the processes had different character-istics on the two sides of the intersection. We do not try to say which part of the process came earlier and which later.[33] The same is true when we speak of marking. To erase a mark is the exact temporal reverse of imposing a mark; to speak of im-posing or erasing is to presuppose a temporal direction. In many cases, of course, we know on other grounds that certain kinds of interactions are irreversible. Light filters absorb some frequencies, so that they transform white light into red. Filters do not furnish missing frequencies to turn red light into white. But until we have gone into the details of the physics of irreversible processes, it is best to think of causal interactions in temporally symmetric terms, and to take the causal connections furnished by causal processes as symmetric connections. Causal processes and causal interactions do not furnish temporal asymmetry; conjunctive forks fulfill that func-tion.

It has been mentioned [. . .] that the cause C in an interactive fork does not *statistically* screen off the effect A from the effect B. There is, however, a kind of *causal* screening off that is a feature of macroscopic interactive forks.[34] In order for ABC to form an interactive fork, there must be causal processes connecting C to A and C to B. Suppose we mark the process that connects C to A at some point between C and A. If we do not specify the relations of temporal priority, then the mark may be transmitted to A or it may be transmitted to C. But whichever way it goes, the mark will not be transmitted to the process connecting C to B. Similarly, a mark imposed upon the process connecting C to B will not be transmitted to the process connecting C to A. This means that no causal influence can be transmitted from A to B or from B to A via C. C constitutes an effective causal barrier between A and B even if A and B exhibit the sort of statistical correlation formulated in (9). The same kind of causal screening occurs in the conjunctive fork, of course, but in forks of this type C statistically screens off A from B as well.

It may be worth noting that marks can be transmitted through interactions, so that if A, C, and B have a linear causal order, a mark made in the process connecting A with C may be transmitted through C to B. For example, if A stands for the impulsion of a cue ball by a cue stick, C for the collision of the cue ball with the 8-ball, and B for the cue ball dropping into the pocket, then it may happen that the tip of the cue stick leaves a blue chalk mark on the cue ball, and that this mark remains on the cue ball until it drops into the pocket. More elaborate examples are commonplace. If a disk jockey at a radio station plays a record that has a scratch on its surface, the mark will persist through all of the many physical processes that take place between the contact of the stylus with the scratch and the click that is perceived by someone who is listening to that particular station.

Perfect Forks

In dealing with conjunctive and interactive forks, it is advisable to restrict our attention to the cases in which $P(A|C)$ and $P(B|C)$ do not assume either of the extreme values of zero or one. The main reason is that the relation

(10) $P(A.B|C) = P(A|C) \times P(B|C) = 1$

may represent a limiting case of either a conjunctive or an interactive fork, even though (10) is a special case of equation (1) and it violates relation (9).

Consider the Annie Oakley example once more. Suppose that she returns to the special shooting gallery time after time. Given that practice makes perfect (at least in her case), she improves her skill until she can invariably hit the knife-edge in the manner that results in the two fragments finding their respective bull's-eyes. Up until the moment that she has perfected her technique, the results of her trials exemplified interactive forks. It would be absurd to claim that when she achieves perfection, the splitting of the bullet no longer constitutes a causal interaction, but must now be regarded as a conjunctive fork. The essence of the interactive fork is to achieve a high correlation between two results; if the correlation is perfect, we can ask no more. It is, one might say, an arithmetical accident that when perfection occurs, equation (1) is fulfilled while the inequality (9) must be violated. If probability values were normalized to some value other than 1, that result would not obtain. It therefore seems best to treat this special case as a third type of fork—the *perfect fork*.

Conjunctive forks also yield perfect forks in the limit. Consider the example of illness due to consumption of poisonous mushrooms. If we assume—what is by no means always the case—that anyone who consumes a significant amount of the mushrooms in question is certain to become violently ill, then we have another instance of a perfect fork. Even when these limiting values obtain, however, there is still no direct interaction between the processes leading respectively to the two cases of severe gastrointestinal distress.

The main point to be made concerning perfect forks is that when the probabilities take on the limiting values, it is impossible to tell from the statistical relationships alone whether the fork should be considered interactive or conjunctive. The fact that relations (1)–(4), which are used in the characterization of conjunctive forks, are satisfied does not constitute a sufficient basis for making a judgment about the temporal orientation of the fork. Only if we can establish, on separate grounds, that the perfect fork is a limiting case of a conjunctive (rather than an interactive) fork, can we conclude that the event at the vertex is a common cause rather than a common effect.[35] Perfect forks need to be distinguished from the other two types mainly to guard against this possible source of confusion.

The Causal Structure of the World

In everyday life, when we talk about cause-effect relations, we think typically (though not necessarily invariably) of situations in which one event (which we call the cause) is linked to another event (which we call the effect) by means of a causal process. Each of the two events in this relation is an interaction between two (or more) intersecting processes. We say, for example, that the window was broken by boys playing baseball. In this situation, there is a collision of a bat with a ball (an interactive fork), the motion of the ball through space (a causal process), and a collision of the ball with the window (an interactive fork). We say, for another example, that turning a switch makes the light go on. In this case, an interaction between a switching mechanism and an electrical circuit leads to a process consisting of a motion of electric charges in some wires, which in turn leads to emission of light from a filament. Homicide by shooting provides still another example. An interaction between a gun and a cartridge propels a bullet (a causal process) from the gun to the victim, where the bullet then interacts with the body of the victim.

The foregoing characterization of causal processes and various kinds of causal forks provides, I believe, a basis for understanding three fundamental aspects of causality:

> 1. *Causal processes* are the means by which structure and order are *propagated* or transmitted from one space-time region of the universe to other times and places.
> 2. *Causal interactions*, as explicated in terms of interactive forks, constitute the means by which *modifications in structure* (which are propagated by causal processes) are *produced*.
> 3. Conjunctive *common causes*—as characterized in terms of conjunctive forks— play a vital role in the *production* of structure and order. In the conjunctive fork, it will be recalled, two or more processes, which are physically independent of one another and which do not interact directly with each other, arise out of some special set of background conditions. The fact that such special background conditions exist is the source of a correlation among the various effects that would be utterly improbable in the absence of the common causal background.

There is a striking difference between conjunctive common causes on the one hand and causal processes and interactions on the other. Causal processes and causal interactions seem to be governed by the basic laws of nature in ways that do not apply to conjunctive forks. Consider two paradigms of causal processes, namely, an electromagnetic wave propagating through a vacuum and a material particle moving without any net external forces acting upon it. Barring any causal interactions in both cases, the electromagnetic wave is governed by Maxwell's equations and the material particle is governed by Maxwell's first law of motion (or its counterpart in relativity theory). Causal interactions are typified by various sorts of collisions. The correlations between the changes that occur in the processes involved are governed— in most, if not all, cases—by fundamental physical conservation laws.

Conjunctive common causes are not nearly as closely tied to the laws of nature. It should hardly require mention that to the extent that conjunctive forks involve causal processes and causal interactions, the laws of nature apply as sketched in the preceding paragraph. However, in contrast to causal processes and causal interactions, conjunctive forks depend crucially upon de facto background conditions. Recall some of the examples mentioned previously. In the plagiarism example, it is a nonlawful fact that two members of the same class happen to have access to the same file of term papers. In the mushroom poisoning example, it is a non-lawful fact that the players sup together out of a common pot. In the twin quasar example, it is a de facto condition that the quasar and the elliptical galaxy are situated in such a way that light coming to us from two different directions arises from a source that radiates quite uniformly from extended portions of its surface.

There is a close parallel between what has just been said about conjunctive forks and what philosophers like Reichenbach (1956, chapter 3) and Grünbaum (1973, chapter 8) have said about entropy and the second law of thermodynamics. Consider the simplest sort of example. Suppose we have a box with two compartments that are connected by a window that can be opened or closed. The box contains equal numbers of nitrogen (N_2) and oxygen (O_2) molecules. The window is open, and all of the N_2 molecules are in the left-hand compartment, while all of the O_2 molecules are in the right-hand compartment. Suppose that there are two molecules of each type. If they are distributed randomly, there is a probability of $2^{-4} = 1/16$ that they would be segregated in just that way—a somewhat improbable coincidence.[36] If there are five molecules of each type, the chance of finding all of the N_2 molecules in the left compartment and all of the O_2 molecules in the right is a bit less than $1/1000$— fairly improbable. If the box contains fifty molecule of each type, the probability of the same sort of segregation would equal $2^{-100} \simeq 10^{-30}$—extremely improbable. If the box contains Avogadro's number of molecules—forget it! In a case of this sort, we would conclude without hesitation that the system had been prepared by closing the window that separates the two compartments, and by filling each compartment separately with its respective gas. The window must have been opened just prior to our examination of the box. What would be a hopelessly improbable coincidence if attributed to chance is explained straightforwardly on the supposition that separate supplies of each of the gases were available beforehand. The explanation depends upon an antecedent state of the world that displays de facto orderliness.

Reichenbach generalized this point in his *hypothesis of the branch structure* (1956, section 16). It articulates the manner in which new sorts of order arise from preexisting states of order. In the thermodynamic context, we say that low entropy states (highly ordered states) do not emerge spontaneously in isolated systems, but, rather, they are produced through the exploitation of the available energy in the immediate environment. Given the fundamentality and ubiquity of entropy considerations, the foregoing parallel suggests that the conjunctive fork also has basic physical significance. If we wonder about the original source of order in the world, which makes possible both the kind of order we find in systems in states of low entropy and the kind of order that we get from conjunctive forks, we must ask the cosmologist how and why the universe evolved into a state characterized by vast supplies of available energy. It does not seem plausible to suppose that order can emerge except from de facto prior order.

In dealing with the interactive fork, I defined it in terms of the special case in which two processes come into an intersection and two processes emerge from it. The space-time diagram has the shape of an x. It does not really matter whether, strictly speaking, the processes that come out of the intersection are the same as the processes that entered. In the case of Compton scattering, for instance, it does not matter whether we say that the incident photon was scattered, with a change in frequency and a loss of energy, or say that one photon impinged upon the electron and another photon emerged. With trivial revisions, the principle CI can accommodate either description.

There are, of course, many other forms that interactions may exhibit. In the Annie Oakley example, two processes (the bullet and the knife-edge) enter the interaction, but three processes (the two bullet fragments and the knife-edge) emerge. We must be prepared to deal with even more complicated cases, but that should present no difficulty, for the basic features of causal interactions can be seen quite easily in terms of the x-type.

Two simpler kinds of interactions deserve at least brief mention. The first of these consists of two processes that come together and fuse into a single outgoing process. Because of the shape of the space-time diagram, I shall call this a λ-type interaction. As one example, consider a snake and a mouse as two distinct processes that merge into one as the snake ingests and digests the mouse. A hydrogen atom, which absorbs a photon and then exists for a time in an excited state, furnishes another.

The other simple interaction involves a single process that bifurcates into two processes. The shape of the space-time diagram suggests that we designate it a y-type interaction. An amoeba that divides to form two daughter amoebas illustrates this sort of interaction. A hydrogen atom in an excited state, which emits a photon in decaying to the ground state, provides another instance.

Since a large number of fundamental physical interactions are of the y-type or the λ-type (see, e.g., Feynman, 1962, or Davies, 1979) there would appear to be a significant advantage in defining interactive forks in terms of these configurations, instead of the x-type. Unfortunately, I have not seen how this can be accomplished, for it seems essential to have two processes going in and two processes coming out in order to exploit the idea of mutual modification. I would be more than pleased if someone could show how to explicate the concept of causal interaction in terms of these simpler types.

Concluding Remarks

There has been considerable controversy since Hume's time regarding the question of whether causes must precede their effects, or whether causes and effects might be simultaneous with each other. It seems to me that the foregoing discussion provides a reasonable resolution of this controversy. If we are talking about the typical cause-effect situation, which I characterized previously in terms of a causal process joining two distinct interactions, then we are dealing with cases in which the cause must precede the effect, for causal propagation over a finite time interval is an essential feature of cases of this type. If, however, we are dealing simply with a causal interaction—an intersection of two or more processes that produces lasting changes in each of them—then we have simultaneity, since each process intersects the other at the same time. Thus it is the intersection of the white light pulse with the red filter that produces the red light, and the light becomes red at the very time of its passage through the filter. Basically, propagation involves lapse of time, while interaction exhibits the relation of simultaneity.

Another traditional dispute has centered upon the question of whether statements about causal relations pertain to individual events, or whether they hold properly only with respect to classes of events. Again, I believe, the foregoing account furnishes a straightforward answer. I have argued that causal processes, in many instances, constitute the causal connections between cause and effect. A causal process is an individual entity, and such entities transmit causal influence. An individual process can sustain a causal connection between an individual cause and an individual effect. Statements about such relations need not be construed as disguised generalizations. At the same time, it should be noted, we have used statistical relations to characterize conjunctive forks. Thus, strictly speaking, when we invoke something like the principle of the common cause, we are implicitly making assertions involving statistical generalizations. Causal relations, it seems to me, have both particular and general aspects.

Throughout the discussion of causality [. . .] I have laid particular stress upon the role of causal processes, and I have even suggested the abandonment of the so-called event ontology. It might be asked whether it would not be possible to carry through the same analysis, within the framework of an event ontology, by considering processes as continuous series of events. I see no reason for supposing that this program could not be carried through, but I would be inclined to ask why we should bother to do so. One important source of difficulty for Hume, if I understand him, is that he tried to account for causal connections between noncontiguous events by interpolating intervening events. This approach seemed only to raise precisely the same questions about causal connections between events, for one had to ask how the causal influence is transmitted from one intervening event to another along the chain. As I argued in [part 2], the difficulty can be circumvented if we look to processes to provide the causal connections. Focusing upon processes rather than events has, in my opinion, enormous heuristic (if not systematic) value. As John Venn said in 1866, "Substitute for the time honoured 'chain of causation,' so often introduced into discussions upon this subject, the phrase a 'rope of causation,' and see what a very different aspect the question will wear" (Venn, 1866, p. 320).

Notes

1. Carnap (1950), section 44.
2. Ibid., sections 50–51.
3. Salmon (1968). Because of this difference with Carnap—that is, my claim that inductive logic requires rules of acceptance for the purpose of establishing statistical generalizations—I do not have the thoroughgoing "pragmatic" or "instrumentalist" view of science Hempel attributes to Richard Jeffrey and associates with Carnap's general conception of inductive logic. Cf. Hempel (1962), pp. 156–63.
4. Salmon (1967), pp. 90–95.
5. See Hempel (1965a), sections 3.2–3.3. In the present essay I am not at all concerned with explanations of the type Hempel calls "deductive-statistical." For greater specificity, what I am calling "statistical explanation" might be called "statistical-relevance explanation," or "S-R explanation" as a handy abbreviation to distinguish it from Hempel's D-N, D-S, and I-S types.
6. Hempel (1962), section 3.
7. Rudolf Carnap (1949).
8. Reichenbach (1949), section 69.
9. Hempel (1962), section 13, and (1965a), section 3.6. Here, Hempel says, "Nonconjunctiveness presents itself as an inevitable aspect of [inductive-statistical explanation],

and thus as one of the fundamental characteristics that set I-S explanation apart from its deductive counterparts."

10. See Nelson Goodman (1965), chapter III. I have suggested a resolution in (Salmon 1963) pp. 252–61.

11. Reichenbach (1956) chapter IV.

12. I find the attempts of Harré and Madden (1975) and Wright (1976) to evade this issue utterly unconvincing. It will be evident [. . .] that the problems of explicating such concepts as causal connections, causal interactions, and cause-effect relations cannot be set aside as mere philosophical quibbles.

13. It might be objected that the alternation of night with day, and perhaps Kepler's 'laws,' do not constitute genuine lawful regularities. This consideration does not really affect the present argument, for there are plenty of regularities, lawful and nonlawful, that do not have explanatory force, but that stand in need of causal explanation.

14. Indeed, in Italian, there is one word, *perchè*, which means both "why" and "because." In interrogative sentences it means "why" and in indicative sentences it means "because." No confusion is engendered as a result of the fact that Italian lacks two distinct words.

15. In this latter work (1948), regrettably, Russell felt compelled to relinquish empiricisim. I shall attempt to avoid such extreme measures.

16. See (Mackie, 1974) for an excellent historical and systematic survey of the various approaches.

17. (Rothman, 1960) contains a lively discussion of pseudo-processes.

18. For example, our discussion of the Minkowski light cone made reference to paths of possible light rays; such a path is one that would be taken by a light pulse if it were emitted from a given space-time point in a given direction. Special relativity seems to be permeated with reference to possible light rays and possible causal connections, and these involve counterfactuals quite directly. See (Salmon, 1976) for further elaborations of this issue, not only with respect to special relativity but also in relation to other domains of physics. A strong case can be made, I believe, for the thesis that counterfactuals are scientifically indispensable.

19. Zeno's arrow paradox and its resolution by means of the 'at-at' theory of motion are discussed in (Salmon, 1975; 2nd ed., 1980, chapter 2). Relevant writings by Bergson and Russell are reprinted in (Salmon, 1970a); the introduction to this anthology also contains a discussion of the arrow paradox.

20. The probabilities that appear in these formulas must, I think, be construed as physical probabilities—that is, as frequencies or propensities. In [Salmon, 1984] I give my reasons for rejecting the propensity interpretation; hence, I construe them as frequencies. Thus I take the variables $A, B, C, \ldots$ which appear in the probability expressions to range over classes.

21. Reichenbach's proof goes as follows. By the theorem on total probability, we may write;

(a) $$P(A.B) = P(C) \times P(A.B|C) + P(\bar{C}) \times P(A.B|\bar{C})$$

(b) $$P(A) = P(C) \times P(A|C) + P(\bar{C}) \times P(A|\bar{C})$$

(c) $$P(B) = P(C) \times P(B|C) + P(\bar{C}) \times P(B|\bar{C}).$$

By virtue of equations (1) and (2), (a) can be rewritten:

(d) $$P(A.B) = P(C) \times P(A|C) \times P(B|C) + P(\bar{C}) \times P(A|\bar{C}) \times P(B|\bar{C}).$$

Now (b), (c), and (d) can be combined to yield:

(e) $$P(A.B) - P(A) \times P(B) = P(C) \times P(A|C) \times P(B|C) + P(\bar{C})$$
$$\times P(A|\bar{C}) \times P(B|\bar{C}) - [P(C) \times P(A|C) + P(\bar{C}) \times P(A|\bar{C})]$$
$$\times [P(C) \times P(B|C) + P(\bar{C}) \times P(B|\bar{C})].$$

Recalling that $P(\bar{C}) = 1 - P(C)$, we can, by elementary algebraic operations, transform the right-hand side of (e) into the following form:

(f) $P(C) \times [1 - P(C)] \times [P(A|C) - P(A|\bar{C})] \times [P(B|C) - P(B|\bar{C})]$.

Assuming that $0 < P(C) < 1$, we see immediately from formulas (3) and (4) that (f) is positive. That result concludes the proof that inequality (5) follows from formulas (1)–(4).

22. Because only two effects, A and B, appear in formulas (1)–(4), I mention only two individuals in this example. The definition of the conjunctive fork can obviously be generalized to handle a larger number of cases.

23. If other potential common causes exist, we can form a partition, C_1, C_2, ..., C_n, $\bar{C}$, and the corresponding relations will obtain. Equation (1) would be replaced by

$$P(A.B|C_i) = P(A|C_i) \times P(B|C_i)$$

and equations (3) and (4) would be replaced by

$$P(A|C_i) > P(A|\bar{C})$$
$$P(B|C_i) > P(B|\bar{C}).$$

24. I am assuming that the magnet in one die does not affect the behavior of the other die.

25. This example was offered at a meeting of the Victoria Section of the Australasian Association for History and Philosophy of Science at the University of Melbourne in 1978.

26. As a matter of fact, Crasnow's example, described in the next paragraph, illustrates this point. In that example, we have two events A and B that form a conjunctive fork with an earlier event C, and also with another earlier event C'. Hence, the four events C', A, B, and C form a double conjunctive fork with vertices at C' and C. In this example, C' qualifies as a bona fide common cause, but C is a spurious common cause. Since C precedes A and B, it cannot be a common effect of those two events; nevertheless, the four events fulfill all of the mathematical relations that define conjunctive forks, which shows that double conjunctive forks are logically possible. If it is difficult—or impossible—to find double conjunctive forks that include both a bona fide common cause and a bona fide common effect, the problem is one of physical, rather than mathematical, constraints. The fact that C occurs before A and B rather than after is a physical fact upon which the probability relations in formulas (1)–(4) have no bearing.

27. In (Salmon, 1980), I took Crasnow's example as a counterexample to Reichenbach's theory; but, as Paul Humphreys kindly pointed out in a private communication, this was an error.

28. I am deeply indebted to Philip von Bretzel (1977, note 13) for the valuable suggestion that causal interactions might be explicated in terms of causal forks. For further elaboration of the relations between the two kinds of forks, see (Salmon, 1978.)

29. For an important discussion of the role of energy and momentum transfer in causality, see (Fair, 1979).

30. As explained in (Salmon, 1978), the example of Compton scattering has the advantage of being irreducibly statistical, and thus, not analyzable, even in principle, as a perfect fork (discussed in a subsequent section of this essay).

31. In (Salmon, 1978), I suggested that interactive forks could be defined statistically, in analogy with conjunctive forks, but I now think that the statistical characterization is inadvisable.

32. In an article entitled "When are Probabilistic Explanations Possible?" Suppes and Zanotti begin with the assertion: "The primary criterion of adequacy of a probabilistic causal analysis is that the causal variable should render the simultaneous phenomenological data conditionally independent. The intuition back of this idea is that the common cause of the

phenomena should factor out the observed correlations. So we label the principle the *common cause criterion*" (1981, p. 191, italics in original). This statement amounts to the claim that all common cause explanations involve conjunctive forks; they seem to overlook the possibility that interactive forks may be involved. One could, of course, attempt to defend this principle on the ground that for any interactive cause C, it is possible to find a conjunctive cause C^*. While this argument may be acceptable for macroscopic phenomena, it does not seem plausible for such microscopic phenomena as Compton scattering.

33. The principle CI, as formulated previously, involves temporal commitments of just this sort. However, these can be purged easily by saying that P_1 and P_2 exhibit Q and R, respectively, on one side of the intersection at S, and they exhibit Q' and R', respectively, on the other side of S. With this reformulation, CI becomes temporally symmetric. When one is dealing with questions of temporal anisotropy or 'direction,' this symmetric formulation should be adopted. Problems regarding the structure of time are not of primary concern in this essay; nevertheless, I am trying to develop causal concepts that will fit harmoniously with a causal theory of time.

34. As Bas van Fraassen kindly pointed out at a meeting of the Philosophy of Science Association, the restriction to macroscopic cases is required by the kinds of quantum phenomena that give rise to the Einstein-Podolsky-Rosen problem.

35. It must be an open rather than a closed fork. In suggesting previously that all conjunctive forks have the same temporal orientation, it was to be understood that we were talking about bona fide conjunctive forks, not limiting cases that qualify as perfect forks.

36. Strictly speaking, each of the probabilities mentioned in this example should be doubled, for a distribution consisting of all O_2 molecules in the left and all N_2 molecules in the right would be just as remarkable a form of segregation as that considered in the text. However, it is obvious that a factor of two makes no real difference to the example.

References

Carnap, Rudolf, "Truth and Confirmation," in *Readings in Philosophical Analysis*, edited by H. Feigel and W. Sellars. New York: Appleton-Century-Crofts, 1949, pp. 119–27.

Chaffee, Frederic H., Jr., "The Discovery of a Gravitational Lens." *Scientific American* 243, no. 5 (November 1980), 70–88.

Davies, P. C. W., *The Forces of Nature*. Cambridge: At the University Press, 1979.

Emerson, Ralph Waldo, "Hymn Sung at the Completion of the Battle Monument, Concord," 1836.

Fair, David, "Causation and the Flow of Energy." *Erkenntnis* 14 (1979), 219–50.

Feynman, Richard, *The Theory of Fundamental Processes*. New York: W. A. Benjamin, 1962.

Goodman, Nelson, *Fact, Fiction, and Forecast*, 2 ed. Indianapolis: Bobbs-Merril, 1965.

Grünbaum, Adolf, *Philosophical Problems of Space and Time*. 2nd ed. Dordrecht: D. Reidel, 1973.

Harré, R., and Madden, E. H., *Causal Powers*. Oxford: Basil Blackwell, 1975.

Hempel, Carl G., "Deductive-Nomological vs. Statistical Explanation." In Herbert Feigel and Grover Maxwell, eds., *Minnesota Studies in the Philosophy of Science*, 3 (1962), 98–169. Minneapolis: University of Minnesota Press.

———, *Aspects of Scientific Explanation and Other Essays in the Philosophy of Science*. New York: Free Press, 1965.

———, "Aspects of Scientific Explanation." In (Hempel, 1965), 331–496.

Hempel, Carl G., and Oppenheim, Paul, "Studies in the Logic of Explanation." *Philosophy of Science* 15 (1948), 135–75; reprinted, with added Postscript, in (Hempel, 1965).

Hume, David, *A Treatise of Human Nature*. Oxford: Clarendon Press, 1888.

————, *An Inquiry Concerning Human Understanding*. Indianapolis, Ind.: Bobbs-Merrill, 1955. Also contains "An Abstract of *A Treatise of Human Nature*.

Mackie, J. L., *The Cement of the Universe*. Oxford: Clarendon Press, 1974.

Reichenbach, Hans, *The Theory of Probability*. Berkeley and Los Angeles: University of California Press, 1949.

————, *The Direction of Time*. Berkeley and Los Angeles: University of California Press, 1956.

Rosen, Deborah A., "An Argument for the Logical Notion of a Memory Trace." *Philosophy of Science* 42 (1975), 1–10.

Rothman, Milton A., "Things That Go Faster Than Light." *Scientific American* 203, no. 1 (July 1960), 142–52.

Russell, Bertrand, *The Analysis of Matter*. London: George Allen and Unwin, 1927.

————, *Mysticism and Logic*. New York: W. W. Norton, 1929.

————, *Human Knowledge, Its Scope and Limits*. New York: Simon and Schuster, 1948.

Salmon, Wesley C., "On Vindicating Induction." *Philosophy of Science* 30 (1963), 252–61. Also published in Henry E. Kyburg, Jr., and Ernest Nagel, eds., *Induction: Some Current Issues*. Middletown, Conn.: Wesleyan University Press.

————, *The Foundations of Scientific Inference*. Pittsburgh: University of Pittsburgh Press, 1967.

————, "Who needs inductive acceptance rules?" In Imre Lakatos, ed., *The Problem of Inductive Logic*. Amsterdam: North-Holland, 1968, 139–44.

————, *Zeno's Paradoxes*. Indianapolis, Ind.: Bobbs-Merrill, 1970a.

————, *Space, Time, and Motion: A Philosophical Introduction*. Encino, Calif: Dickenson, 1975. 2nd ed., Minneapolis: University of Minnesota Press, 1980.

————, "Foreword." In Hans Reichenbach, *Laws, Modalities, and Counterfactuals*, vii–xlii. Berkeley/Los Angeles/London: University of California Press, 1976.

————, "Why ask, 'Why?'?" *Proceedings and Addresses of the American Philosophical Association* 51 (1978), 683–705. Reprinted in (Salmon, 1979a).

————, "Probabilistic Causality." *Pacific Philosophical Quarterly* 61 (1980), 50–74.

Sayre, Kenneth M., "Statistical Models of Causal Relations." *Philosophy of Science* 44 (1977), 203–14.

Suppes, Patrick, *A Probabilistic Theory of Causality*. Amsterdam: North-Holland, 1970.

Suppes, Patrick, and Zanotti, Mario, "When are Probabilistic Explanations Possible?" *Synthese* 48 (1981), 191–99.

van Fraassen, Bas C., *The Scientific Image*. Oxford: Clarendon Press, 1980.

Venn, John, *The Logic of Chance*. London: Macmillan, 1866.

von Bretzel, Philip, "Concerning a Probabilistic Theory of Causation Adequate for the Causal Theory of Time." *Synthese* 35 (1977), 173–90. Reprinted in (Salmon, 1979a).

Winnie, John, "The Causal Theory of Space-time." In John Earman, Clark Glymour, and John Stachel, eds., *Minnesota Studies in the Philosophy of Science*, 8 (1977), 134–205. Minneapolis: University of Minnesota Press.

Wright, Larry, *Teleological Explanation*. Berkeley and Los Angeles: University of California Press, 1976.

5

A Deductive-Nomological Model of Probabilistic Explanation

Peter Railton

What if some things happen by chance—can they nonetheless be explained? How?

Some things *do* happen by chance, according to the dominant interpretation of our present physical theory, the probabilistic interpretation of quantum mechanics. Nonetheless, they can be explained: by that theory, in virtually the same way as deterministic phenomena—deductive-nomologically. At least, that is what I hope to show in this essay.

Our universe may not be deterministic, but all is not chaos. It is governed by laws of two kinds: probabilistic (such as the laws concerning barrier penetration and certain other quantum phenomena) and non-probabilistic (such as the laws of conservation of mass-energy, charge, momentum, etc.).[1] Were the probabilism of laws of the first sort remediable by suitable elaboration of laws of the second sort, the universe would be deterministic after all, and the problem of explaining chance phenomena would no longer be with us. However, indications are that physical indeterminism is irremediable, and that the universe exhibits not only chances, but lawful chances. I will argue that we come to understand chance phenomena, even when the chance involved is extremely remote, by subsuming them under these irremediably probabilistic laws.

1. Introductory Remarks on Explanation

Do I offer a deductive-nomological (D-N) model of probabilistic explanation because I believe that nomic subsumption always explains?—No. There are familiar-enough kinds of non-explanatory D-N arguments, for example, those that deduce the explanandum from nomically-related symptoms or after-the-fact conditions alone, citing no causes.

Yet it will not do simply to add to the D-N model a requirement that the explanans contain causes whenever the explanandum is a particular fact. First, some particular facts may be explained non-causally, for example, by subsumption under

Reprinted from *Philsophy of Science*, 45, (1978) pp. 206–226. Copyright © 1978 by the Philosophy of Science Association. With permission of the Philosophy of Science Association and the author.

structural laws such as the Pauli exclusion principle. Second, even where causal explanation is called for, the existence of general, causal laws that cover the explanandum has not always been sufficient for explanation: the search for explanation has also taken the form of a search for mechanisms that underlie these laws. 'Mechanisms,' however, is not meant to suggest a parochial attitude toward the nomic connections—deterministic or otherwise—that tie the world together and make explanation possible.

An example may help clarify the notion of mechanism appealed to here. The following D-N argument suffices to forecast *that* nasty weather lies ahead, but not to explain *why* this is so:

(S) The glass is falling.
Whenever the glass falls the weather turns bad.

The weather will turn bad. ([5], p. 106)

Now nothing works like a barometer for predicting the weather, but nothing like a barometer works for changing it. So it is often maintained that (S) lacks explanatory efficacy because barometers lack the appropriate causal efficacy. The following inference, then, remedies the lack of the first because "it proves that the fact is a fact by citing causes and not mere symptoms" ([5], p. 107):

(C) The glass is falling.
Whenever the glass is falling the atmospheric pressure is falling.
Whenever the atmospheric pressure is falling the weather turns bad.

The weather will turn bad. ([5], p. 106)

Yet as explanations go, (C) is also lacking: we remain in the dark as to *why* the weather will turn bad. No connection between cause and effect, no mechanism by which falling atmospheric pressure produces a change for the worse in the weather, has been revealed. I do not doubt that some account of this mechanism exists; my point is that its existence is what makes (C) superior to (S) for explanatory purposes.

(C), if moderated by boundary conditions and put less qualitatively, would supply us the capability to predict *and* control the weather (whenever, as in a laboratory simulator, we can manipulate the atmospheric pressure). While prediction and control may exhaust our practical problems in the natural world, the unsatisfactoriness of (C) shows that explanation is an activity not wholly practical in purpose. The goal of understanding the world is a theoretical goal, and if the world is a machine—a vast arrangement of nomic connections—then our theory ought to give us some insight into the structure and workings of the mechanism, above and beyond the capability of predicting and controlling its outcomes. Until supplemented with an account of the nomic links connecting changes in atmospheric pressure to changes in the weather, (C) will explain but poorly. Knowing enough to subsume an event under the right kind of laws is not, therefore, tantamount to knowing the *how* or *why* of it. As the explanatory inadequacies of successful practical disciplines remind us: explanations must be more than potentially-predictive inferences or law-invoking recipes.

Is the deductive-nomological model of explanation therefore unacceptable?— No, just incomplete. Calling for an account of the mechanism leaves open the nature of that account, and as far as I can see, the model explanations offered in scientific texts are D-N when complete, D-N sketches when not. What is being urged is that D-N explanations making use of true, general, causal laws may legitimately be regarded as unsatisfactory unless we can back them up with an account of the mechanism(s) at work. "An account of the mechanism(s)" is a vague notion, and one obviously admitting of degrees of thoroughness, but I will not have much to say here by way of demystification. If one sees what is lacking in (C)—a characterization, whether sketchy or blow-by-blow, of how it is that declining atmospheric pressure effects the changes we describe as "a worsening of the weather," that is, a more or less complete filling-in of the links in the causal chains—one has the rough idea.

The D-N probabilistic explanations to be given below do not explain by giving a deductive argument terminating in the explanandum, for it will be a matter of chance, resisting all but *ex post facto* demonstration. Rather, these explanations subsume a fact in the sense of giving a D-N account of the chance mechanism responsible for it, and showing that our theory implies the existence of some physical possibility, however small, that this mechanism will produce the explanandum in the circumstances given. I hope the remarks just made about the importance of revealing mechanisms have eased the way for an account of probabilistic explanation that focuses on the indeterministic mechanisms at work, rather than the "nomic expectability" of the explanandum.

2. Hempel's Inductive-Statistical Model

For Hempel, a statistical explanation (what is called elsewhere in this paper 'a probabilistic explanation') is one that "makes essential use of at least one law or theoretical principle of statistical form" ([3], p. 380). Since Hempel distinguishes between statistical laws and mere statistical generalizations, and asserts that the former apply only where "peculiar, namely probabilistic, modes of connection" exist among the phenomena ([3], p. 377), his characterization permits statistical explanation only of genuinely indeterministic processes.[2] Were some process to have the appearance of indeterminism owing to arcane workings or uncontrolled initial conditions, then no "peculiar . . . probabilistic" modes of connection would figure essentially in explaining this "pseudo-random" process's outcomes. Not only would statistical explanation be unnecessary for such a process, it would be impossible: no probabilistic *laws* would govern it.

For example, it has been observed that 99% of all cases of infectious mononucleosis involve lymph-gland swelling. The exceptions might be due to a process that randomly misfires 1% of the time. Or, they might arise from the operation of an unknown deterministic mechanism that works to inhibit swelling whenever a patient begins in a particular initial condition, which as a mere matter of fact is typical of 1% of the population. If initial conditions could be partitioned into two mutually exclusive and jointly exhaustive classes S and $-S$, such that all Ss by law eventually develop swelling, and all $-S$s do not, the generalization "99% of all cases of infectious mononucleosis develop lymph-gland swelling" would have been shown to be no law, but merely a descriptive report of observed relative frequencies. No law, it cannot support a statistical explanation. But discovering it not to be a law is just discovering that statistical explanation is uncalled for, since each case of mononucleosis will have been of type S or type $-S$ from the outset.

On the other hand, suppose that no such partition of initial conditions exists. Then the presence or absence of swelling is presumably due to a "peculiar . . . probabilistic" connection between disease and symptom, that is, a real causal indeterminism with probability .99 in each case to produce swelling. The generalization in question would thus be nomological, creating both the possibility and the necessity of statistical explanation.

Given such genuine statistical laws, how does Hempel claim statistical explanation should proceed? He begins his account by distinguishing two sorts of statistical explanation. The first, *deductive-statistical (D-S)* explanation, involves "the deductive subsumption of a narrow statistical uniformity under more comprehensive ones" ([3], p. 380). The second, he argues, is of a qualitatively different sort:

> Ultimately . . . statistical laws are meant to be applied to particular occurrences and to establish explanatory and predictive connections among them. ([3], p. 381)

To make such laws relevant to "particular occurrences," Hempel believes we must go beyond the reach of deduction, and so he proposes an inductive model of statistical explanation.

Inductive-statistical (I-S) explanation proceeds by adducing statistical laws and associated initial conditions relative to which the explanandum is highly probable. High relative probability is required because, on Hempel's view, statistical laws become explanatorily relevant to an individual chance event only by giving us a basis upon which to inductively infer its occurrence with "practical certainty." Yet although an I-S explanation shows the explanandum to have been "nomically expectable" relative to the explanans, it does not permit detachment of a conclusion; it is less an inference than the expression of an inferential relationship: the explanandum receives a high degree of epistemic support from the explanans. If, for example, we learn that Jones has contracted infectious mononucleosis, we may infer with practical certainty that he will develop lymph-gland swelling. The same inference serves as an I-S explanation of the swelling, should it occur. Should it not occur, we would have no explanation for *this,* on Hempel's model.

However, further investigation of Jones' medical history might reveal that he suffered mononucleosis once before, and failed to develop any swelling. Let us suppose that such individuals have a much higher than normal probability of *not* showing swelling in any later bouts with mononucleosis, say .9 rather than .01. This new law and new information about Jones together permit an inference with practical certainty to the conclusion that he will *not* develop swelling, and thus support a corresponding I-S explanation. Relative to these new facts, however, no I-S explanation would be available should Jones, improbably, develop swelling. What are we to say now about the previous I-S explanation, which had just the opposite result? Hempel would reject it as no longer *maximally specific* relative to what we believe about Jones' case. The requirement of maximal specificity is a complicated affair,[3] but the basic idea is that we refer each case to the narrowest class of cases to which our present beliefs assign it in which the explanandum has a characteristically different probability. In Jones' case, the narrower class is clearly the class of those contracting mononucleosis for a second time who failed to develop lymph-gland swelling the first time.

If more information about Jones or new discoveries about mononucleosis turn up, we may be forced to move on to still another explanation. I-S explanations must be relativized to our current "epistemic situation," and are subject to change along with it. Hempel notes that this sets off I-S explanations from D-N and D-S explanations in a fundamental way:

> . . . *the concept of statistical explanation for particular events is essentially relative to a given knowledge situation as represented by a class K of accepted statements.* . . . [W]e can significantly speak of true D-N and D-S explanations: they are those potential D-N and D-S explanations whose premises (and hence also conclusions) are true—no matter whether this happens to be known or believed, and thus no matter whether the premises are included in *K*. But this idea has no significant analogue for I-S explanation. . . . ([3], pp. 402–3)

On Hempel's view, neither of the two contradictory explanations concerning Jones contains false premises, and the explananda in each case do indeed receive the degree of support indicated. It is just that we no longer regard the evidential relationship expressed by the first as explanatorily relevant. Were Jones to develop swelling after all, it would now have to be regarded as inexplicable.

What I take to be the two most bothersome features of I-S arguments as models for statistical explanation—the requirement of high probability and the explicit relativization to our present epistemic situation (bringing with it an exclusion of questions about the truth of I-S explanations)—derive from the inductive character of such inferences, not from the nature of statistical explanation itself. If a non-inductive model for the statistical explanation of particular facts is given, there need be no temptation to require high probability or exclude truth.

3. Jeffrey's Criticism of I-S Explanation

Richard C. Jeffrey has criticized Hempel's account on the grounds that statistical explanation is not a form of inference at all, except when the probability of the explanandum is "so high as to allow us to reason, in *any* decision problem, as if its probability were 1" ([5], p. 105). For such exceptional, "beautiful" cases, Jeffrey accepts I-S inferences as explanatory because they provide virtual "proof that the phenomenon *does* take place" ([5], p. 106).

For unbeautiful cases, there is no way of proving (in advance) that the explanandum phenomenon will occur. According to Jeffrey, the explanation *why* such unbeauties come to be is a curt "By chance." He has more to say on *how* they come about:

> . . . in the statistical case I find it strained to speak of knowledge *why* the outcome is such-and-such. I would rather speak of *understanding the process*, for the explanation is the same no matter what the outcome: it consists of a statement that the process is a stochastic one, following such-and-such a law.[4] ([5], p. 24)

Jeffrey is surely right, as against Hempel, that probable and improbable outcomes of indeterministic processes are equally explicable, and explicable in the same way. After all, why should it be explicable that a genuinely random wheel of fortune with 99 red stops and 1 black stop came to a halt on red, but inexplicable that it halted on black? Worse, on Hempel's view, halting at any *particular* stop would be inexplicable, even though the wheel must halt at some particular stop in order to yield the explicable outcome *red*.

But I fail to see how Jeffrey can defend his exemption of beautiful cases against a similar line of argument. If the burden in statistical explanation really lies with *"understanding the process . . .* no matter what the outcome," then why should it matter whether the outcome is so highly probable "as to allow us to reason, in *any* decision problem, as if its probability were 1?" The neglect Jeffrey shows here toward minute chances is appropriate for the practical task of decision-making (and perhaps explained by his generally subjectivist approach to probability), but we must not overlook them in the not-entirely-practical task of explaining. Virtually impossible events may occur, and they deserve and can receive the same explanation as the merely improbable or the virtually certain.

4. A D-N Model of Probabilistic Explanation

I will present my account of probabilistic explanation by developing an example of just such "practically negligible"—but physically real and lawful—chance: alpha-decay in long-lived radioactive elements. The mean-life of the more stable radio-nuclides is so long as to make the probability for any particular nucleus of such an element to decay during our lifetimes effectively zero. But our nuclear theory shows that it is *not* zero, and explains how such rarities can occur.

On the account offered here, probabilistic explanations will be either true or false independent of our epistemic situation. Moreover, to explain, they must be true. Here I am following Hempel's usage in calling an explanatory argument *true* just in case it is valid and its premises are true ([3], p. 338). Such an explanation will *not* be true if the probabilistic laws it invokes are not true; in particular, it will not be true unless the process responsible for the explanandum is genuinely inde-terministic. If alpha-decay is to serve as our paradigm for probabilistic explanation, we must be correct in assuming that the probabilistic wave-mechanical account of particle transmission through the nuclear potential barrier tells us all there is to know about the cause of alpha-decay. At least, it must be true that there are no hidden variables characterizing unknown initial conditions that suffice to account for alpha-decay deterministically. However, I take it to be uncontroversial that alpha-decay is an indeterministic process, if any is.

Let us suppose that we are given an individual instance of alpha-decay to explain: a nucleus of radionuclide uranium^{238}, call it 'u', has emitted an alpha-particle during the time interval lasting from t_0 to $t_0 + \theta$, where θ is very small and expressed in standard units. Since the mean-life of U^{238} is 6.5×10^9 years, the probability of observing a decay by u during this interval is exceedingly small, but unquestionably exists (witness the decay). This probability can be given precisely by using the ra-dioactive decay constant λ_{238} characteristic of all atoms of U^{238}. Significantly, we need not know when in the course of the history of u time t_0 occurs: the probability of decay is unaffected by the age of the atom. Therefore, as long as decay has not yet occurred, individual "trials"—consisting of observing a single isolated radio-active nucleus for successive intervals of the same length—are statistically indepen-dent. Using these two facts we can determine the probability of decay for individual nuclei during any time interval chosen: it will be 1 minus the probability that any such nucleus *survives* the interval intact; for u, $(1 - \exp(-\lambda_{238} \cdot \theta))$.

To obtain experimental confirmation of this value, we infer *from* the probability to decay of individual nuclei *to* statistical features of sample populations of nuclei, for example, half-life and mean-life. These predicted statistical features are then checked against actual observed relative frequencies in large populations over long intervals. *Physical* probabilities of the sort being considered here are therefore to be contrasted with *statistical* probabilities; the former express the strength of a certain physical possibility for a given system, while the latter reduce to claims about the (limiting) relative frequencies of traits in sample populations. Much well-founded doubt has been expressed about the applicability of statistical probabilities to single cases, but physical probabilities are *located* in the features of the single case. Therefore, we can understand our nuclear theory as implying strictly universal (physical) probability-attributing laws of the form:

(1)   All nuclei of radioelement E have probability $(1 - \exp(-\lambda_E \cdot t))$ to emit an alpha-particle during any time interval of length t, unless subjected to environmental radiation.

Because schema (1) is universal in form, its instances are candidates for law premises in deductive-nomological inferences concerning individual nuclei. Thus, for u:

(2)   (a)   All nuclei of U^{238} have probability $(1 - \exp(-\lambda_{238} \cdot \theta))$ to emit an alpha-particle during any interval of length θ, unless subjected to environmental radiation.

(b) u was a nucleus of U^{238} at time t_0, and was subjected to no environmental radiation before or during the interval $t_0 - (t_0 + \theta)$.

(c) u had probability $(1 - \exp(-\lambda_{238} \cdot \theta))$ to emit an alpha-particle during the interval $t_0 - (t_0 + \theta)$.

(2), it appears, gives a D-N explanation only of the fact that u had such-and-such a probability to decay during the interval in question, but we should look a bit closer. I submit that (2), when supplemented as follows, is the probabilistic explanation of u's decay:

(3)   A derivation of (2a) from our theoretical account of the mechanism at work in alpha-decay.
The D-N inference (2).
A parenthetic addendum to the effect that u did alpha-decay during the interval $t_0 - (t_0 + \theta)$.

Am I merely making a virtue of necessity, and saying that since (3) contains all we can say about u's decay, (3) must explain it? In fact, there is a great deal more we could say about u's decay. Deliberately left out of (3) are innumerable details about the experimental apparatus (temperature, pressure, location, etc.), about the beliefs and expectations of those monitoring the experiment, and about the epistemic position of the scientific community at the time. These facts are omitted as *explanatorily irrelevant* to u's decay because they are *causally irrelevant* to the physical possibility for decay that obtained during the interval in question, and to whether or not that possibility was realized.[5] A full account of these notions of explanatory and causal relevance is not possible here, so instead I will go on to argue that what (3) comprises *is* explanatorily relevant, and explanatory.

I must begin this task with a defense of the nomological status of (2a), and of the legitimacy of treating it as a covering law for u's decay. The following criterion of nomologicality will be used: a law is a universal truth derivable from our theory without appeal to particular facts. This criterion of course lacks generality (what counts as theory if not the laws themselves?), fails to segregate natural from logical laws, picks out only so-called "universal" (as opposed to "local") laws, and is entirely too vague (how to distinguish "particular facts" from the rest?). But I trust it will do for now. The motive for excluding "particular facts" is that some true, universal statements derivable from our theory *plus* particular facts would not normally be regarded as universal laws, but would at best be "local laws," for example, "All *Homo neanderthalensis* live during the late Pleistocene age."

The generalization in question here, (2a), is derived by solving the Schrödinger wave equation for an alpha-particle of energy ≈ 4.2 MeV for the potential regions in and around the nucleus of an element with atomic number 92 and atomic weight 238, none of which are "particular facts," plus some simplifying assumptions about the structure of the nucleus and the distinctness of the alpha-particle within it prior to decay. While it is forbidden by classical physics for a low-energy particle like the ≈ 4.2 MeV alpha-particle associated with U^{238} to pass through the 24.2 MeV potential barrier surrounding so massive a nucleus, the quantum theory predicts that the probability amplitude for finding such an alpha-particle outside the potential barrier is non-zero. Thus a transmission coefficient for U^{238} alpha-particles is determined, which, given certain simplifying assumptions about the goings-on inside the nucleus, yields the probability that such a particle will tunnel out of the potential well "per unit time for one nucleus," namely, λ_{238} ([1], p. 175). (2a) thus neither reports a summary of past observations nor expresses a mere statistical uniformity that scattered initial conditions would lead us to anticipate. Instead, it is a law of irreducibly probabilistic form, assigning definite, physically determined probabilities to individual systems.

It follows that the derivation of conclusions from (2a) by universal instantiation and *modus ponens* is unexceptionable. Were (2a) but a statistical generalization, properly understood as meaning "$(1 - \exp(-\lambda_{238} \cdot \theta))N$ of U^{238} nuclei in samples of sufficiently large size N, on average, decay during the interval $t_0 - (t_0 + \theta)$," it could not undergo universal instantiation, and would not permit detachment of a conclusion about the probability obtaining in a single case.

Further, if the wave equation does indeed tell us all there is to know about the mechanism involved in nuclear barrier penetration, it follows that nothing more can be said to explain why the observed decay of u took place, once we have shown how (2a) is derived from our account of this mechanism, and established that (2) is valid and that (3)'s parenthetic addendum is true.

Still, does (3) explain why the decay took place? It does not explain why the decay *had* to take place, nor does it explain why the decay *could be expected to* take place. And a good thing, too: there is no *had to* or *could be expected to* about the decay to explain—it is not only a chance event, but a very improbable one. (3) does explain why the decay *improbably* took place, which is how it did. (3) accomplishes this by demonstrating that there existed at the time a small but definite physical possibility of decay, and noting that, by chance, this possibility was realized. The derivation of (2a) that begins (3) shows, by assimilating alpha-decay to the chance process of potential barrier tunneling, how this possibility comes to exist. If alpha-decays are chance phenomena of the sort described, then once our theory has achieved all that (3) involves, it has explained them to the hilt, however unsettling this may be to *a priori* intuitions. To insist upon stricter subsumption of the explanandum is not merely to demand what (alas) cannot be, but what decidedly should not be: sufficient reason that one probability rather than another be realized, that is, chances without chance.

Because of the peculiar nature of chance phenomena, it is explanatorily relevant whether the probability in question was realized, even though there is no before-the-fact explanatory *argument*, deductive or inductive, to this conclusion. Indeed, it is the absence of such an argument that makes a place in probabilistic explanation for a parenthetic addendum concerning whether the possibility became actual in the circumstances given. These addenda may offend those who believe that explanations must always be arguments, but at the most general level explanations are *accounts*, not arguments. It so happens that for deterministic phenomena inferences of a particular kind—D-N arguments meeting the desiderata suggested in section 1—*are* explanatory accounts, and this for good reasons. However, indeterministic phenomena are a different matter, and explanatory accounts of them must be different as well. If the present model is accepted, then almost all of the explanatory burden in probabilistic explanation can be placed on deductive arguments—those characterizing the indeterministic mechanism and those attributing a certain probability to the explanandum. But these arguments leave out a crucial part of the story: did the chance fact obtain?

The parenthetic addendum fills this gap in the account, and communicates information that is relevant to the causal origin of the explanandum by telling us that it came about as the realization of a particular physical possibility. Further, it permits us to chain probabilistic explanations together to make more comprehensive explanations, in which each link is able to bear the full explanatory burden for the fact it covers, and is capable of leading us on to the next fact in the causal sequence being explained. From (2) alone we cannot move directly to an account of what the alpha-particle did to a nearby photographic plate, but only to a probability (and a miserably low one) that this account will be true. The parenthetic addendum to (3) furnishes a non-probabilistic premise from which to begin an account of the condition of the photographic plate: the occurrence of an alpha-decay in the vicinity. Dropping off the addendum leaves an explanation, but it is a D-N explanation of the occurrence of a particular probability, not a probabilistic explanation of the occurrence of a particular decay.

The scheme for probabilistic explanation of particular chance facts by nomic subsumption that is being offered here, the *deductive-nomological-probabilistic* (D-N-P) model, is this. First we display (or truthfully claim an ability to display) a derivation from our theory of a law of essentially probabilistic form, complete with an account of how the law applies to the deterministic process in question. The derived law is of the form:

(4a) $$\forall t \forall \chi [F_{\chi,t} \rightarrow \text{Prob}(G)_{\chi,t} = p]$$

"At any time, anything that is F has probability p to be G."

Next, we adduce the relevant fact(s) about the case at hand, e:

(4b) F_{e,t_0}

"e is F at time t_0,"

and draw the obvious conclusion:

(4c) $\text{Prob}(G)_{e,t_0} = p$

"e has probability p to be G at time t_0."

To which we add parenthetically, and according to how things turn out:

(4d) $(G_{e,t_0}/-G_{e,t_0})$

"(e did/did not become G at t_0)."

Whether a D-N-P explanation is true will depend solely upon the truth-values of its premises and addendum, and the validity of its logic. I leave open what becomes of a D-N-P explanation that contains true laws, initial facts, and addendum, but botches the theoretical account of the laws invoked. Let us simply say that the more botched, the less satisfactory the explanation.

The law premise (4a) will be true if all things at all times satisfy the conditional '$F_{x,t} \rightarrow \text{Prob}(G)_{x,t} = p$', using whatever reading of '$\rightarrow$' we decide upon for the analysis of natural laws in general. It will be false if there exists a partition of the Fs into those with *physical* probability r to be G and those with *physical* probability s to be G, where $s \neq r \neq p$. Such a partition might exist according to some *other* interpretation of probability, but this would not affect the truth of (4a). For example, suppose that a coin toss meeting certain specifications is an indeterministic event with probability $1/2$ of yielding heads. We now perform the experiment of repeating such a toss a great many times. Curiously, all and only even numbered tosses yield heads. This result supplies certain frequentists with grounds for saying that Prob(heads, even-numbered toss) = 1, while Prob(heads, odd-numbered toss) = 0.[6] But because all tosses met the specification laid down, the probabiity of heads was the same, $1/2$, on each toss, despite the curious behavior. Such behavior may make us suspicious of our original claims about the indeterminacy of the process or about the physical probability it has of producing heads, but is no proof against them. Indeed, the original probability attribution requires us to assign a definite physical probability to just such an untoward sequence of outcomes, the occurrence of which therefore hardly contradicts this attribution.

The particular fact premise (4b) will be true iff e is an F during the time in question, and not either an F^* (with probability $r \neq p$ to be G) or an F^{**} (with probability $q = p$ to be G, but unlike an F in other respects). Using the (let us say) true law that all F^{**}s have probability $q = p$ to be G, and the falsehood that e is an F^{**}, we could derive a true conclusion, indistinguishable from (4c). Hence the requirement that the *premises* be true if the argument is to explain; and if we reason logically from true premises, the conclusion will take care of itself.

5. Epistemic Relativity and Maximal Specificity Disowned

Have I kept my promise to give an account of probabilistic explanation free from relativization to our present epistemic situation?

Let us return to explanation (3), and admit that it is not the whole story: 23% of the alpha-particles emitted by U^{238} have kinetic energy 4.13 MeV, while the remaining 77% have 4.18 MeV. Therefore there are two different decay constants, $\lambda_{238}^{4.13}$ and $\lambda_{238}^{4.18}$; both are distinct from λ_{238}, used in (3). Hence we must be quite careful in stating what exactly (3) explains. It does *not* explain the particular *event* observed, for this was either a 4.13 or a 4.18 MeV decay, neither of which has probability λ_{238} in unit time. Instead, (3) explains the particular *fact about* the event observed that we set out to explain, namely, that an alpha-decay with unspecified energy (or direction, or angular momentum, etc.) took place at nucleus u during the time interval in question. This fact *does* have probability λ_{238} of obtaining in unit time, representing the sum of the two energy-correlated probabilities with which such a decay might occur.

If we should learn that the decay of u was of a 4.18 MeV alpha-particle, an explanation of *this* fact would have to be referred to the more specific class of decays with probability $\lambda_{238}^{4.18}$ in unit time. Is the maximal specificity requirement thereby resurrected? There is no need for it. (3) is not an unspecific explanation of this more specific fact, but a fallacious one. It would be logically corrupt to conclude from law (2a) that an individual U^{238} nucleus has probability $(1 - \exp(-\lambda_{238} \cdot \theta))$ to decay *with energy 4.18 MeV* during any interval of length θ, since (2a) says nothing whatsoever about decay energies. The only relevant conclusion to draw from (2a) is (2c), which remains true in the face of our more detailed knowledge about the event in question. Nor is law (2a) falsified by the discovery of a 23:77 proportional distribution of decay energies, and the associated difference in decay rates. For according to our nuclear theory, there is no difference in initial condition between a nucleus about to emit a 4.13 MeV alpha-particle and one about to emit a 4.18 MeV alpha-particle. It remains true that *all* U^{238} nuclei have probability λ_{238} to decay in unit time, but it is further true that all have probability $\lambda_{238}^{4.13}$ to decay one way, and probability $\lambda_{238}^{4.18}$ to decay another.

It must next be determined whether the existence of a difference in probability *due to* a difference in initial condition can be handled by the D-N-P model without appeal to a maximal specificity requirement. To permit consideration of possible epistemological complications, it will be assumed that neither the difference in probability nor the partition of initial conditions is known at the start.

Imagine that, although we do not know it, in virtue of certain permanent structural features 23% of all naturally-occurring U^{238} nuclei fall into a class P, and the remaining 77% into a class $-P$, such that only those in P have any probability of emitting a 4.13 MeV alpha-particle, and only those in class $-P$ have any probability of emitting a 4.18 MeV alpha-particle. Suppose further that these two laws have been derived:

(5) (a) All U^{238} nuclei of type P have probability $(1 - \exp(-\lambda_{238}^{4.13} \cdot t))$ to emit a 4.13 MeV alpha-particle during any time interval of length t, unless subjected to environmental radiation.

(b) All U^{238} nuclei of type $-P$ have probability $(1 - \exp(-\lambda_{238}^{4.18} \cdot t))$ to emit a 4.18 MeV alpha-particle during any time interval of length t, unless subjected to environmental radiation.

Note that, by our assumptions, the specification of the kinetic energy of the particle (possibly) emitted may be dropped from (5a) and (5b) without altering the truth of either.

Until the structural differences between types P and $-P$ are discovered and understood, (3) will stand as the accepted explanation of u's decay. However, once (5a) and (5b) have become known, it will be clear from the fact that u's alpha-emission had kinetic energy 4.18 MeV that u must have been of type $-P$ prior to decay. Thus a more specific account of u's decay will be available to scientists, who, already familiar with the theoretical derivation of law (5b), offer the following truncated D-N-P version of this account:

(6) (a) All nuclei of U^{238} of type $-P$ have probability $(1 - \exp(-\lambda_{238}^{4.18} \cdot \theta))$ to emit an alpha-particle during any time interval of length θ, unless subjected to environmental radiation.

     (b) u was a nucleus of U^{238} of type $-P$ at t_0, and was subjected to no environmental radiation before or during the interval $t_0 - (t_0 + \theta)$.

     (c) u had probability $(1 - \exp(-\lambda_{238}^{4.18} \cdot \theta))$ to emit an alpha-particle during the interval $t_0 - (t_0 + \theta)$.

     (d) (And it did.)

On the Hempelian model (modified so as to permit I-S explanations of improbable phenomena), there is no problem in accounting for the previous acceptability of the I-S counterpart of (3), or for its present unacceptability. (3) had been maximally specific relative to our previous beliefs about alpha-decay in U^{238}, but no longer is, and so is superseded by the more specific (relatively speaking) I-S counterpart of (6).

On the D-N-P model, too, there is no problem in accounting for the acceptability of (3) prior to the discovery of class $-P$ and law (5b): (3)'s premises (and, of course, addendum) were taken to be true. The question is whether, in light of current beliefs, (3) can be ruled out—and (6) ruled in—without invocation of Hempelian constraints. Resolution of the problem (3) and (6) pose through epistemic relativization and maximal specificity requirements seems to me unacceptable. If we were to attribute to nucleus u two unequal probabilities to alpha-decay in a specified way during a single time interval, adding, "Let's pick the most specifically defined value for explanatory purposes," we'd be showing an unseemly tolerance for contradiction in our nuclear theory—and why stop there? Better face up to the confrontation over truth between (3) and (6), and replace complex and unappealingly relativistic maximal specificity requirements with the simple requirement of truth. The D-N-P model does this. The current unacceptability of (3) is located not in premises insufficiently specific, but in premises insufficiently true, that is, false. Contrary to (3)'s purported covering law (2a), not all nuclei of U^{238} have probability λ_{238} to decay in unit time if unperturbed by radiation—in fact, none do. In spite of giving accurate expectation values for decay rates in large samples of U^{238}, (2a) is false, and so explanation (3) is ruled out as unsound. Explanation (6), on the other hand, meets the simple requirement of truth, and rules itself in.[7]

Problems about incomplete, misleading, or false beliefs do not bear on whether D-N-P explanations have unrelativized truth-values, but concern rather difficulties in *establishing* the truth-values they unrelativistically have. Relativization to our current epistemic situation comes into play only when we begin to discuss whether a given D-N-P explanation *seems* true. Whether it *is* true is another matter.

6. Objections to the D-N-P Model

I cannot pretend to have said enough about deductive-nomological-probabilistic explanation to have characterized this model adequately. Such reservations as were expressed in section 1 about taking nomic subsumption under a causal law as sufficient for explanation are still in force, and little has been done—except by way of example—to show how the account offered here might accommodate them.

That the probabilistic laws invoked in D-N-P explanations are even (in some relevant sense) *causal* cannot be defended until a plausible account of physical probability has been worked out, a task well beyond the scope of this paper. Under a *propensity interpretation,* probability has the characteristics sought: a probability is the expression of the strength of a physical tendency in an individual chance system to produce a particular outcome; it is therefore straightforwardly applicable to single cases; and it is (in a relevant sense) causally responsible for that outcome whenever it is realized. However, propensities are notoriously unclear. For now I can at best assume that clarification is possible, point to a promising start in the attempt to do so—R. N. Giere, "Objective, Single-Case Probabilities and the Foundations of Statistics" ([2])—, and admit that the D-N-P model is viable only if sense can be made of propensities, or of objective, physical, lawful, single-case probabilities by any other name.

As for the requirement that explanations elucidate mechanisms, I can only repeat that an essential role is played in D-N-P explanations by the theoretical deduction of the probabilistic law(s) covering the explanandum.

In lieu of further exposition, I offer the beginnings of a defense, hoping thereby to sketch out the account a bit more fully in those areas most likely to be controversial.

Because It Applies Only to Genuinely Indeterministic Processes, of Which There Are Few (If Any), D-N-P *Explanation Is Too Restricted in Scope.*

It is widely believed that the probabilities associated with standard gambling devices, classical thermodynamics, actuarial tables, weather forecasting, etc., arise not from any underlying physical indeterminism, but from an unknown or uncontrolled scatter of initial conditions. If this is right, then D-N-P explanation would be inapplicable to these phenomena even though they are among the most familiar objects of probabilistic explanation. I do not, however, find this troublesome: if something does not happen by chance, it cannot be explained by chance. The use of epistemic or statistical probabilities in connection with such phenomena unquestionably has instrumental value, and should not be given up. What must be given up is the idea that *explanations* can be based on probabilities that have no role in bringing the world's explananda about, but serve only to describe deterministic phenomena.[8]

Whether there *are* any probabilities that enter into the mechanisms of nature is still debated, but the successes of the quantum-mechanical formalism, and the existence of "no hidden variable" results for it, place the burden of proof on those who would insist that physical chance is an illusion.

It could be objected more justly that D-N-P explanation is too broad, not too narrow, in scope. Once restrictions have been lifted from the value a chance may have in probabilistic explanation, virtually all explanations of particular facts must become probabilistic. All but the most basic regularities of the universe stand forever in peril of being interrupted or upset by intrusion of the effects of random processes. It might seem a fine explanation for a light's going out that we opened the only circuit connecting it with an electrical power source, but an element of chance was involved: had enough atoms in the vicinity of the light undergone spontaneous beta-decay at the right moment, the electrons emitted could have kept it glowing. The success of a social revolution might appear to be explained by its overwhelming popular support, but this is to overlook the revolutionaries' luck: if all the naturally unstable nuclides on earth had commenced spontaneous nuclear fission in rapid succession, the triumph of the people would never have come to pass.

No doubt this proliferation of probabilistic explanations is counterintuitive, but contemporary science will not let us get away with any other sort of explanation in these cases—it simply cannot supply the requisite non-probabilistic laws. Because they figure in the way things *work* tiny probabilities appropriately figure in explanations of the way things *are,* even though they scarcely ever show up in the way things turn out.

The D-N-P *Model Breaks the Link Between Prediction and Explanation.*

Hempel has justified a "qualified thesis of the structural identity of explanation and prediction" with this principle:

> Any rationally acceptable answer to the question "Why did X occur?" must offer information which shows that X was to be expected—if not definitely, then at least with reasonable certainty. ([3], pp. 367–368)

Abundantly many D-N-P explanations—all those covering less than highly probable facts—violate this condition.

However, to abide by this condition and renounce explanations with meager probabilities I take to be worse. Why forgo the explanations of improbable phenomena offered by our theories, when these explanations provide as much of an account of why (and how) their explananda occur as do the explanations of "reasonably certain" phenomena that Hempel's condition sanctions?

Too restrictive as it stands, Hempel's condition may be taken in a way not incompatible with D-N-P explanation. A D-N-P explanation does yield one prediction that is perfectly strict, to the effect that a certain physical probability exists in the circumstances given. If this probability fails to obtain, or to have the value attributed to it, the explanation must be false. It is a complaint against the world, not against the D-N-P model, that a direct, non-statistical test for the presence or value of this probability may prove impossible. Remarkably, the mechanisms of the world leave room for spontaneous nuclear disintegrations. Equally remarkably, our physical theory gives us insight into how they come about, and assigns determinate probabilities to them. These probabilities are connected to the rest of our theory by laws that permit both prediction and (where means exist) control: if undisturbed, nucleus a will have probability p to alpha-decay (so we should expect a's decay with *epistemic* probability p); and if we wish to alter p, our theory tells us how a must be disturbed.

It has been objected to the view of probability taken in this paper that unless probability attributions are interpreted as predictions about how relative frequencies will *actually* come out in the long run, probabilistic laws lack empirical content. Thus if the relative frequency of decayed atoms in a large sample of some radioelement were, over a great length of time, to diverge significantly from the probability theoretically attributed to decay, that attribution would not be "borne out," that is, would be falsified.Otherwise, it is argued, probabilistic laws are compatible with all frequencies, and empirically vacuous.

But it is impossible for a world to "bear out" all of its probability-attributing laws in this sense. For these laws imply, among other things, that it is extremely unlikely that *all* actual long-run sequences will show a relative frequency near to the single-case probability. Therefore, the demand that all long-run decay rates nearly match all corresponding decay constants comes to a demand that nothing improbable show up in the long run, which is itself an improbability showing up in the long run. Intended to clear things up on the epistemological front, this proposal cannot even get out of its own way.

By Splitting Apart Probabilistic Explanation and Induction, the D-N-P Model Has Lost the Point of Probabilistic Explanation.

Behind this objection lies the view that probabilistic (or statistical) explanation is an activity fundamentally unlike D-N explanation. A probabilistic explanation is seen as a piece of detective work. Unable to give a causal demonstration of the explanandum from evidence thus far assembled, we develop hypotheses, which are judged by how probable they are on the evidence, and whether they make the explanandum sufficiently probable. In the end, we put forward the most convincing inductive argument yet found—the one making the explanandum most antecedently probable, given what else we know about events leading up to it.

This view of probabilistic explanation confuses epistemic with objective probability, and induction with explanation. Perhaps responsible for this confusion is the similarity of the tasks of explaining a phenomenon, gathering support for such an explanation, and gathering before- or after-the-fact evidence for a phenomenon's occurrence. This confusion is abetted by misleading ways of talking about "strong" or "good" explanations. We should distinguish the following, (i) *A strong (good) explanation* is one that has great theoretical power, regardless of how well-confirmed it is or how probable it holds the exlpanandum to be. (ii) *A strong (good) candidate for explanation* is a proffered explanation with well-confirmed *premises,* regardless of how probable it holds the explanandum to be and irrespective of how theoretically powerful it happens to be. (iii) *A strong (good) reason for believing that the explanandum fact will obtain* is furnished by before-the-fact evidence that leads, via one's theory, to an expectation of the explanandum with high epistemic probability. (iv) *A strong (good) reason for believing that the explanandum fact obtained* is given by any evidence that lends high epistemic probability to the proposition that the explanandum fact is a fact. Strong after-the-fact evidence, even for very improbable events, may be easy to come by. Reasons of types (iii) and (iv) need have nothing to do with explanation, and may be based on symptoms (Will it rain today?—Harry's rheumatism is acting up) or even less causally relevant information (Was Sue upset?—Her brother is certain she would have been).

Although the link between probabilistic explanation and induction is looser on the D-N-P model than on the I-S model, this is no fault: on Hempel's account it was entirely too close. Measuring the strength or "acceptability" of an explanation by the magnitude of the probability it confers on the explanandum blurs the distinctions just made. Keeping (i)–(iv) distinct, the D-N-P model enables us to state quite simply the object of induction in explanation: given a particular fact, to find, and *gather evidence for*, an explanans that subsumes it; given a generalization, to find, and gather evidence for, a higher-level explanans that subsumes it; in all cases, then, to discover and establish a true and relevant explanans. The issue of showing the explanandum to have high (relative or absolute) probability is a red herring, distracting attention from the real issue: the truth or falsity, and applicability, of the laws and facts adduced in explanatory accounts.

Notes

I would like to thank C. G. Hempel, Richard C. Jeffrey, and David Lewis for helpful criticisms of earlier drafts. I am especially indebted to David Lewis for the idea that a propensity interpretation of probability sits best with the account of probabilistic explanation given here. I have greatly benefited from discussions of related matters with Sam Scheffler and David Fair.

1. Let us say rather loosely that a system is deterministic if, for any one instant, its state is physically compatible with only one (not necessarily different) state at each other instant. A system is indeterministic otherwise, but lawfully so if a complete description of its state at some one instant plus all true laws together entail a distribution of probabilities over possible states at later times.

2. Although there is some difficulty in reconciling all that is said in [3] with this conclusion, Hempel now accepts it (personal communication).

3. See, for example, [4].

4. A typographical error has been corrected.

5. Causal relevance is established here via the wave equation. I do not mean to suggest that *causal* relevance is the only explanatory kind; cf. the mention of structural laws in section 1.

Some such notion of causal relevance appears to lie behind Salmon's "statistical-relevance" model of probabilistic explanation. Yet what matters is whether a factor enters into the probabilities present, not the statistics they produce.

6. Cf. the discussion of place selections and homogeneity in [6], sections 4 and 7. Salmon's criterion, which requires formal randomness, would here fail to distinguish a *randomly-produced* regular sequence from a *deterministically-produced* one. Notwithstanding formal similarities, only the latter is appropriately explained non-probabilistically.

7. Explanation (6) is true, however, only under the contrary-to-fact assumption—made for the sake of the example—of the existence of a class $-P$.

8. Of course, we might speak of statistical or epistemic probabilities as causes of, for example, beliefs. But if belief formation is not *physically* probabilistic, then probabilistic

explanation of it would be impossible, in spite of this sort of causal involvement on the part of statistical or epistemic probabilities.

References

[1] Evans, R. D., *The Atomic Nucleus*. New York: McGraw Hill, 1965.
[2] Giere, R. N., "Objective Single-Case Probabilities and the Foundations of Statistics." In *Logic, Methodology and Philosophy of Science,* vol. IV. Edited by P. Suppes, et al. Amsterdam: North-Holland, 1973.
[3] Hempel, C. G., "Aspects of Scientific Explanation." In *Aspects of Scientific Explanation and Other Essays*. New York: Free Press, 1965.
[4] Hempel, C. G., "Maximal Specificity and Lawlikeness in Probabilistic Explanation." *Philosophy of Science* 35 (1968), 116–33.
[5] Jeffrey, R. C., "Statistical Explanation vs. Statistical Inference." In *Essays in Honor of C. G. Hempel*. Edited by N. Rescher et al. Dordrecht: D. Reidel, 1970.
[6] Salmon, W. C., "Statistical Explanation." In *Statistical Explanation and Statistical Relevance*. Edited by W. Salmon. Pittsburgh: University of Pittsburgh Press, 1971.

6

The Pragmatic Theory of Explanation

Bas C. Van Fraassen

1. "The Tower and the Shadow"

During my travels along the Saône and Rhône last year, I spent a day and night at the ancestral home of the Chevalier de St. X . . ., an old friend of my father's. The Chevalier had in fact been the French liaison officer attached to my father's brigade in the first war, which had—if their reminiscences are to be trusted—played a not insignificant part in the battles of the Somme and Marne.

The old gentleman always had *thé à l'Anglaise* on the terrace at five o'clock in the evening, he told me. It was at this meal that a strange incident occurred; though its ramifications were of course not yet perceptible when I heard the Chevalier give his simple explanation of the length of the shadow which encroached upon us there on the terrace. I had just eaten my fifth piece of bread and butter and had begun my third cup of tea when I chanced to look up. In the dying light of that late afternoon, his profile was sharply etched against the granite background of the wall behind him, the great aquiline nose thrust forward and his eyes fixed on some point behind my left shoulder. Not understanding the situation at first, I must admit that to begin with, I was merely fascinated by the sight of that great hooked nose, recalling my father's claim that this had once served as an effective weapon in close combat with a German grenadier. But I was roused from this brown study by the Chevalier's voice.

"The shadow of the tower will soon reach us, and the terrace will turn chilly. I suggest we finish our tea and go inside."

I looked around, and the shadow of the rather curious tower I had earlier noticed in the grounds, had indeed approached to within a yard from my chair. The news rather displeased me, for it was a fine evening; I wished to remonstrate but did not well know how, without overstepping the bounds of hospitality. I exclaimed, "Why must that tower have such a long shadow? This terrace is so pleasant!"

His eyes turned to rest on me. My question had been rhetorical, but he did not take it so.

"As you may already know, one of my ancestors mounted the scaffold with Louis XVI and Marie Antoinette. I had that tower erected in 1930 to mark the exact spot where it is said that he greeted the Queen when she first visited this house, and presented her with a peacock made of soap, then a rare substance. Since the Queen would have been one hundred and seventy-five years old in 1930, had she lived. I had the tower made exactly that many feet high."

It took me a moment to see the relevance of all this. Never quick at sums, I was at first merely puzzled as to why the measurement should have been in feet; but of course I already knew him for an Anglophile. He added drily, "The sun not being alterable in its course, light traveling in straight lines, and the laws of trigonometry being immutable, you will perceive that the length of the shadow is determined by the height of the tower." We rose and went inside.

I was still reading at eleven that evening when there was a knock at my door. Opening it I found the housemaid, dressed in a somewhat old-fashioned black dress and white cap, whom I had perceived hovering in the background on several occasions that day. Courtseying prettily, she asked, "Would the gentleman like to have his bed turned down for the night?"

I stepped aside, not wishing to refuse, but remarked that it was very late—was she kept on duty to such hours? No, indeed, she answered, as she deftly turned my bed covers, but it had occurred to her that some duties might be pleasures as well. In such and similar philosophical reflections we spent a few pleasant hours together, until eventually I mentioned casually how silly it seemed to me that the tower's shadow ruined the terrace for a prolonged, leisurely tea.

At this, her brow clouded. She sat up sharply. "What exactly did he tell you about this?" I replied lightly, repeating the story about Marie Antoinette, which now sounded a bit far-fetched even to my credulous ears.

"The *servants* have a different account," she said with a sneer that was not at all becoming, it seemed to me, on such a young and pretty face. "The truth is quite different, and has nothing to do with ancestors. That tower marks the spot where he killed the maid with whom he had been in love to the point of madness. And the height of the tower? He vowed that shadow would cover the terrace where he first proclaimed his love, with every setting sun—that is why the tower had to be so high."

I took this in but slowly. It is never easy to assimilate unexpected truths about people we think we know—and I have had occasion to notice this again and again.

"Why did he kill her?" I asked finally.

"Because, sir, she dallied with an English brigadier, an overnight guest in this house." With these words she arose, collected her bodice and cap, and faded through the wall beside the doorway.

I left early the next morning, making my excuses as well as I could.

2. A Model for Explanation

I shall now propose a new theory of explanation. An explanation is not the same as a proposition, or an argument, or list of propositions; it is an *answer*. (Analogously, a son is not the same as a man, even if all sons are men, and every man is a son.)

An explanation is an answer to a why-question. So, a theory of explanation must be a theory of why-questions.

To develop this theory, whose elements can all be gleaned, more or less directly, from the preceding discussion, I must first say more about some topics in formal pragmatics (which deals with context-dependence) and in the logic of questions. Both have only recently become active areas in logical research, but there is general agreement on the basic aspects to which I limit the discussion.

2.1. Contexts and Propositions[1]

Logicians have been constructing a series of models of our language, of increasing complexity and sophistication. The phenomena they aim to save are the surface grammar of our assertions and the inference patterns detectable in our arguments. (The distinction between logic and theoretical linguistics is becoming vague, though logicians' interests focus on special parts of our language, and require a less faithful fit to surface grammar, their interests remaining in any case highly theoretical.) Theoretical entities introduced by logicians in their models of language (also called 'formal languages') include domains of discourse ('universes'), possible worlds, accessibility ('relative possibility') relations, facts and propositions, truth-values, and, lately, contexts. As might be guessed, I take it to be part of empiricism to insist that the adequacy of these models does not require all their elements to have counterparts in reality. They will be good if they fit those phenomena to be saved.

Elementary logic courses introduce one to the simplest models, the languages of sentential and quantificational logic which, being the simplest, are of course the most clearly inadequate. Most logic teachers being somewhat defensive about this, many logic students, and other philosophers, have come away with the impression that the oversimplifications make the subject useless. Others, impressed with such uses as elementary logic does have (in elucidating classical mathematics, for example), conclude that we shall not understand natural language until we have seen how it can be regimented so as to fit that simple model of horseshoes and truth tables.

In elementary logic, each sentence corresponds to exactly one proposition, and the truth-value of that sentence depends on whether the proposition in question is true in the actual world. This is also true of such extensions of elementary logic as free logic (in which not all terms need have an actual referent), and normal modal logic (in which non-truth functional connectives appear), and indeed of almost all the logics studied until quite recently.

But, of course, sentences in natural language are typically context-dependent; that is, which proposition a given sentence expresses will vary with the context and occasion of use. This point was made early on by Strawson, and examples are many:

> "I am happy now" is true in context x exactly if the speaker in context x is happy at the time of context x.

where a context of use is an actual occasion, which happened at a definite time and place, and in which are identified the speaker (referent of 'I'), addressee (referent of 'you'), person discussed (referent of 'he'), and so on. That contexts so conceived are idealizations from real contexts is obvious, but the degree of idealization may be decreased in various ways, depending on one's purposes of study, at the cost of greater complexity in the model constructed.

What must the context specify? The answer depends on the sentence being analyzed. If that sentence is

> Twenty years ago it was still possible to prevent the threatened population explosion in that country, but now it is too late

the model will contain a number of factors. First, there is a set of possible worlds, and a set of contexts, with a specification for each context of the world of which it is a part. Then there will be for each world a set of entities that exist in that world, and also various relations of relative possibility among these worlds. In addition there is time, and each context must have a time of occurrence. When we evaluate the above sentence we do so relative to a context and a world. Varying with the context will be the referents of 'that country' and 'now', and perhaps also the relative possibility relation used to interpret 'possible', since the speaker may have intended one of several senses of possibility.

This sort of interpretation of a sentence can be put in a simple general form. We first identify certain entities (mathematical constructs) called propositions, each of which has a truth-value in each possible world. Then we give the context as main task the job of selecting, for each sentence, the proposition it expresses 'in that context'. Assume as simplification that when a sentence contains no indexical terms (like 'I', 'that', 'here', etc.), then all contexts select the same proposition for it. This gives us an easy intuitive handle on what is going on. If A is a sentence in which no indexical terms occur, let us designate as $|A|$ the proposition which it expresses in every context. Then we can generally (though not necessarily always) identify the proposition expressed by any sentence in a given context as the proposition expressed by some indexical-free sentence. For example:

> In context x, "Twenty years ago it was still possible to prevent the population explosion in that country" expresses the proposition "In 1958, it is (tenseless) possible to prevent the population explosion in India"

To give another example, in the context of my present writing, "I am here now" expresses the proposition that Bas van Fraassen is in Vancouver, in July 1978.

This approach has thrown light on some delicate conceptual issues in philosophy of language. Note for example that "I am here" is a sentence which is true no matter what the facts are and no matter what the world is like, and no matter what context of usage we consider. Its truth is ascertainable *a priori*. But the proposition expressed, that van Fraassen is in Vancouver (or whatever else it is) is not at all a necessary one: I might not have been here. Hence, a clear distinction between *a priori* ascertainability and necessity appears.

The context will generally select the proposition expressed by a given sentence A via a selection of referents for the terms, extensions for the predicates, and functions for the functors (that is, syncategorematic words like 'and' or 'most'). But intervening contextual variables may occur at any point in these selections. Among such variables there will be the assumptions taken for granted, theories accepted, world-pictures or paradigms adhered to, in that context. A simple example would be the range of conceivable worlds admitted as possible by the speaker; this variable plays a role in determining the truth-value of his modal statements in that context, relative to the "pragmatic presuppositions." For example, if the actual world is really the only possible world there is (which exists) then the truth-values of modal statements in that context but *tout court* will be very different from their truth-values relative to those pragmatic presuppositions—and only the latter will play a significant role in our understanding of what is being said or argued in that context.

Since such a central role is played by propositions, the family of propositions has to have a fairly complex structure. Here a simplifying hypothesis enters the fray: propositions can be uniquely identified through the worlds in which they are true. This simplifies the model considerably, for it allows us to identify a proposition with a set of possible worlds, namely, the set of worlds in which it is true. It allows the family of propositions to be a complex structure, admitting of interesting operations, while keeping the structure of each individual proposition very simple.

Such simplicity has a cost. Only if the phenomena are simple enough, will simple models fit them. And sometimes, to keep one part of a model simple, we have to complicate another part. In a number of areas in philosophical logic it has already been proposed to discard that simplifying hypothesis, and to give propositions more "internal structure." As will be seen below, problems in the logic of explanation provide further reasons for doing so.

2.2. Questions

We must now look further into the general logic of questions. There are of course a number of approaches; I shall mainly follow that of Nuel Belnap, though without committing myself to the details of his theory.[2]

A theory of questions must needs be based on a theory of propositions, which I shall assume given. A *question* is an abstract entity; it is expressed by an *interrogative* (a piece of language) in the same sense that a proposition is expressed by a declarative sentence. Almost anything can be an appropriate response to a question, in one situation or another; as 'Peccavi' was the reply telegraphed by a British commander in India to the question how the battle was going (he had been sent to attack the province of Sind).[3] But not every response is, properly speaking, an answer. Of course, there are degrees; and one response may be more or less of an answer than another. The first task of a theory of questions is to provide some typology of answers. As an example, consider the following question, and a series of responses:

Can you get to Victoria both by ferry and by plane?

(*a*) Yes.

(*b*) You can get to Victoria both by ferry and by plane.

(*c*) You can get to Victoria by ferry.

(*d*) You can get to Victoria both by ferry and by plane, but the ferry ride is not to be missed.

(*e*) You can certainly get to Victoria by ferry, and that is something not to be missed.

Here (*b*) is the "purest" example of an answer: it gives enough information to answer the question completely, but no more. Hence it is called a *direct answer*. The word 'Yes' (*a*) is a *code* for this answer.

Responses (c) and (d) depart from that direct answer in opposite directions: (c) says properly less than (b)—it is implied by (b)—while (d), which implies (b), says more. Any proposition implied by a direct answer is called a *partial answer* and one which implies a direct answer is a *complete answer*. We must resist the temptation to say that therefore an answer, *tout court,* is any combination of a partial answer with further information, for in that case, every proposition would be an answer to any question. So let us leave (e) unclassified for now, while noting it is still "more of an answer" than such responses as 'Gorilla!' (which is a response given to various questions in the film *Ich bin ein Elephant, Madam,* and hence, I suppose, still more of an answer than some). There may be some quantitative notion in the background (a measure of the extent to which a response really "bears on" the question) or at least a much more complete typology (some more of it is given below), so it is probably better not to try and define the general term "answer" too soon.

The basic notion so far is that of direct answer. In 1958, C. L. Hamblin introduced the thesis that a question is uniquely identifiable through its answers.[4] This can be regarded as a simplifying hypothesis of the sort we come across for propositions, for it would allow us to identify a question with the set of its direct answers. Note that this does not preclude a good deal of complexity in the determination of exactly what question is expressed by a given interrogative. Also, the hypothesis does not identify the question with the disjunction of its direct answers. If that were done, the clearly distinct questions

Is the cat on the mat?
    direct answers: The cat is on the mat.
                         The cat is not on the mat.

Is the theory of relativity true?
    direct answers: The theory of relativity is true.
                         The theory of relativity is not true.

would be the same (identified with the tautology) if the logic of propositions adopted were classical logic. Although this simplifying hypothesis is therefore not to be rejected immediately, and has in fact guided much of the research on questions, it is still advisable to remain somewhat tentative towards it.

Meanwhile we can still use the notion of direct answer to define some basic concepts. One question Q may be said to *contain* another, Q', if Q' is answered as soon as Q is—that is, every complete answer to Q is also a complete answer to Q'. A question is *empty* if all its direct answers are necessarily true, and *foolish* if none of them is even possibly true. A special case is the *dumb* question, which has no direct answers. Here are examples:

1. Did you wear the black hat yesterday or did you wear the white one?
2. Did you wear a hat which is both black and not black, or did you wear one which is both white and not white?
3. What are three distinct examples of primes among the following numbers: 3, 5?

Clearly 3 is dumb and 2 is foolish. If we correspondingly call a necessarily false statement foolish too, we obtain the theorem *Ask a foolish question and get a foolish answer*. This was first proved by Belnap, but attributed by him to an early Indian philosopher mentioned in Plutarch's *Lives* who had the additional distinction of being an early nudist. Note that a foolish question contains all questions, and an empty one is contained in all.

Example 1 is there partly to introduce the question form used in 2, but also partly to introduce the most important semantic concept after that of direct answer, namely presupposition. It is easy to see that the two direct answers to 1 ("I wore the black hat." "I wore the white one") could both be false. If that were so, the respondent would presumably say "Neither," which is an answer not yet captured by our typology. Following Belnap, who clarified this subject completely, let us introduce the relevant concepts as follows:

> a *presupposition*[5] of question Q is any proposition which is implied by all direct answers to Q.
> a *correction* (or *corrective answer*) to Q is any denial of any presupposition of Q.
> the *(basic) presupposition* of Q is the proposition which is true if and only if some direct answer to Q is true.

In this last notion, I presuppose the simplifying hypothesis which identifies a proposition through the set of worlds in which it is true; if that hypothesis is rejected, a more complex definition needs to be given. For example 1, 'the' presupposition is clearly the proposition that the addressee wore either the black hat or the white one. Indeed, in any case in which the number of direct answers is finite, 'the' presupposition is the disjunction of those answers.

Let us return momentarily to the typology of answers. One important family is that of the partial answers (which includes direct and complete answers). A second important family is that of the corrective answer. But there are still more. Suppose the addressee of question 1 answers "I did not wear the white one." This is not even a partial answer, by the definition given: neither direct answer implies it, since she might have worn both hats yesterday, one in the afternoon and one in the evening, say. However, since the questioner is presupposing that she wore at least one of the two, the response is *to him* a complete answer. For the response plus the presupposition together entail the direct answer that she wore the black hat. Let us therefore add:

> a *relatively complete answer* to Q is any proposition which, together with the presupposition of Q, implies some direct answer to Q.

We can generalize this still further: a complete answer to Q, relative to theory T, is something which together with T, implies some direct answer to Q—and so forth. The important point is, I think, that we should regard the introduced typology of answers as open-ended, to be extended as needs be when specific sorts of questions are studied.

Finally, which question is expressed by a given interrogative? This is highly context-dependent, in part because all the usual indexical terms appear in interrogatives. If I say "Which one do you want?" the context determines a range of objects over which my 'which one' ranges—for example, the set of apples in the basket on my arm. If we adopt the simplifying hypothesis discussed above, then the main task of the context is to delineate the set of direct answers. In the 'elementary questions' of Belnap's theory ('whether-questions' and 'which-questions') this set of direct answers is specified through two factors: a *set of alternatives* (called the *subject* of the question) and *request* for a selection among these alternatives and, possibly, for certain information about the selection made ('distinctness and completeness claims'). What those two factors are may not be made explicit in the words used to frame the interrogative, but the context has to determine them exactly if it is to yield an interpretation of those words as expressing a unique question.

2.3. A Theory of Why-Questions

There are several respects in which why-questions introduce genuinely new elements into the theory of questions.[6] Let us focus first on the determination of exactly what question is asked, that is, the contextual specification of factors needed to understand a why-interrogative. After that is done (a task which ends with the delineation of the set of direct answers) and as an independent enterprise, we must turn to the evaluation of those answers as good or better. This evaluation proceeds with reference to the part of science accepted as "background theory" in that context.

As [an] example, consider the question "Why is this conductor warped?" The questioner implies that the conductor is warped, and is asking for a reason. Let us call the proposition that the conductor is warped the *topic* of the question (following Henry Leonard's terminology, 'topic of concern'). Next, this question has a *contrast-class,* as we saw, that is, a set of alternatives. I shall take this contrast-class, call it X, to be a class of propositions which includes the topic. For this particular interrogative, the contrast could be that it is *this* conductor rather than *that* one, or that this conductor has warped rather than retained its shape. If the question is "Why does this material burn yellow" the contrast-class could be the set of propositions: this material burned (with a flame of) colour x.

Finally, there is the respect-in-which a reason is requested, which determines what shall count as a possible explanatory factor, the relation of *explanatory relevance*. In the first example, the request might be *for events "leading up to" the warping*. That allows as relevant an account of human error, of switches being closed or moisture condensing in those switches, even spells cast by witches (since the evaluation of what is a good answer comes later). On the other hand, the events leading up to the warping might be well known, in which case the request is likely to be for the standing conditions that made it possible for those events to lead to this warping: the presence of a magnetic field of a certain strength, say. Finally, it might already be known, or considered immaterial, exactly how the warping is produced, and the question (possibly based on a misunderstanding) may be about exactly what function this warping fulfils in the operation of the power station. Compare "Why does the blood circulate through the body?" answered (1) "because the heart pumps the blood through the arteries" and (2) "to bring oxygen to every part of the body tissue."

In a given context, several questions agreeing in topic but differing in contrast-class, or conversely, may conceivably differ further in what counts as explanatorily relevant. Hence we cannot properly ask what is relevant to this topic, or what is relevant to this contrast-class. Instead we must say of a given proposition that it is or is not relevant (in this context) to the topic with respect to that contrast-class. For example, in the same context one might be curious about the circumstances that led

Adam to eat the apple rather than the pear (Eve offered him an apple) and also about the motives that led him to eat it rather than refuse it. What is 'kept constant' or 'taken as given' (that he ate the fruit; that what he did, he did to the apple) which is to say, the contrast-class, is not to be dissociated entirely from the respect-in-which we want a reason.

Summing up then, the why-question Q expressed by an interrogative in a given context will be determined by three factors:

> The *topic* P_k
> The *contrast-class* $X = \{P_1, \ldots, P_k, \ldots\}$
> The *relevance relation R*

and, in a preliminary way, we may identify the abstract why-question with the triple consisting of these three:

$$Q = \langle P_k, X, R \rangle$$

A proposition A is called *relevant to Q* exactly if A bears relation R to the couple $\langle P_k, X \rangle$.

We must now define what are the direct answers to this question. As a beginning let us inspect the form of words that will express such an answer:

$$(*) \; P_k \; in \; contrast \; to \; \text{(the rest of)} \; X \; because \; A$$

This sentence must express a proposition. What proposition it expresses, however, depends on the same context that selected Q as the proposition expressed by the corresponding interrogative ("Why P_k?"). So some of the same contextual factors, and specifically R, may appear in the determination of the proposition expressed by (*).

What is claimed in answer (*)? First of all, that P_k is true. Secondly, (*) claims that the other members of the contrast-class are not true. So much is surely conveyed already by the question—it does not make sense to ask why Peter rather than Paul has paresis if they both have it. Thirdly, (*) says that A is true. And finally, there is that word 'because': (*) claims that A is a *reason*.

This fourth point we have awaited with bated breath. Is this not where the inextricably modal or counterfactual element comes in? But not at all; in my opinion, the word 'because' here signifies only that A is relevant, in this context, to this question. Hence the claim is merely that A bears relation R to $\langle P_k, X \rangle$. For example, suppose you ask why I got up at seven o'clock this morning, and I say "because I was woken up by the clatter the milkman made." In that case I have interpreted your question as asking for a sort of reason that at least includes events-leading-up-to my getting out of bed, and my word 'because' indicates that the milkman's clatter was that sort of reason, that is, one of the events in what Salmon would call the causal process. Contrast this with the case in which I construe your request as being specifically for a motive. In that case I would have answered "No reason, really. I could easily have stayed in bed, for I don't particularly want to do anything today. But the milkman's clatter had woken me up, and I just got up from force of habit

I suppose." In this case, I do not say 'because' for the milkman's clatter does not belong to the relevant range of events, as I understand your question.

It may be objected that 'because A' does not only indicate that A is *a* reason, but that it is *the* reason, or at least that it is a good reason. I think that this point can be accommodated in two ways. The first is that the relevance relation, which specifies what sort of thing is being requested as answer, may be construed quite strongly: "give me a motive strong enough to account for murder," "give me a statistically relevant preceding event not screened off by other events," "give me a common cause," etc. In that case the claim that the proposition expressed by A falls in the relevant range, is already a claim that it provides a telling reason. But more likely, I think, the request need not be construed that strongly; the point is rather that anyone who answers a question is in some sense tacitly claiming to be giving a good answer. In either case, the determination of whether the answer is indeed good, or telling, or better than other answers that might have been given, must still be carried out, and I shall discuss that under the heading of 'evaluation'.

As a matter of regimentation I propose that we count (*) as a direct answer *only if A is relevant.*[7] In that case, we don't have to understand the claim that A is relevant as explicit part of the answer either, but may regard the word 'because' solely as a linguistic signal that the words uttered are intended to provide an answer to the why-question just asked. (There is, as always, the tacit claim of the respondent that what he is giving is a good, and hence a relevant answer—we just do not need to make this claim part of the answer.) The definition is then:

B is a *direct answer* to question $Q = \langle P_k, X, R \rangle$ exactly if there is some proposition A such that A bears relation R to $\langle P_k, X \rangle$ and B is the proposition which is true exactly if (P_k; and for all $i \neq k$, not P_i; and A) is true

where, as before, $X = \{P_1, \ldots, P_k, \ldots\}$. Given this proposed definition of the direct answer, what does a why-question presuppose? Using Belnap's general definition we deduce:

> a why-question *presupposes* exactly that
>
> (a) its topic is true
> (b) in its contrast-class, only its topic is true
> (c) at least one of the propositions that bears its relevance relation to its topic
> and contrast-class, is also true.

However, as we shall see, if all three of these presuppositions are true, the question may still not have a *telling* answer.

Before turning to the evaluation of answers, however, we must consider one related topic: when does a why-question arise? In the general theory of questions, the following were equated: question Q arises, all the presuppositions of Q are true. The former means that Q is not to be rejected as mistaken, the latter that Q has some true answer.

In the case of why-questions, we evaluate answers in the light of accepted background theory (as well as background information) and it seems to me that this drives a wedge between the two concepts. Of course, sometimes we reject a why-question because we think that it has no true answer. But as long as we do not think that, the question does arise, and is not mistaken, regardless of what is true.

To make this precise, and to simplify further discussion, let us introduce two more special terms. In the above definition of 'direct answer', let us call proposition A the *core* of answer B (since the answer can be abbreviated to '*Because A*'), and let us call the proposition that (P_k *and for all* $i \neq k$, *not* P_i) the *central presupposition* of question Q. Finally, if proposition A is relevant to $\langle P_k, X \rangle$ let us also call it relevant to Q.

In the context in which the question is posed, there is a certain body K of accepted background theory and factual information. This is a factor in the context, since it depends on who the questioner and audience are. It is this background which determines whether or not the question arises; hence a question may arise (or conversely, be rightly rejected) in one context and not in another.

To begin, whether or not the question genuinely *arises*, depends on whether or not K implies the central presupposition. As long as the central presupposition is not part of what is assumed or agreed to in this context, the why-question does not arise at all.

Secondly, Q presupposes *in addition* that one of the propositions A, relevant to its topic and contrast-class, is true. Perhaps K does not imply that. In this case, the question will still arise, provided K does not imply that all those propositions are false.

So I propose that we use the phrase "the question arises in this context" to mean exactly this: K implies the central presupposition, and K does not imply the denial of any presupposition. Notice that this is very different from "all the presuppositions are true," and we may emphasize this difference by saying "arises in context." The reason we must draw this distinction is that K may not tell us which of the possible answers is true, but this *lacuna* in K clearly does not eliminate the question.

2.4. *Evaluation of Answers*

The main problems of the philosophical theory of explanation are to account for legitimate rejections of explanation requests, and for the asymmetries of explanation. These problems are successfully solved, in my opinion, by the theory of why-questions as developed so far.

But that theory is not yet complete, since it does not tell us how answers are evaluated as telling, good, or better. I shall try to give an account of this too, and show along the way how much of the work by previous writers on explanation is best regarded as addressed to this very point. But I must emphasize, first, that this section is not meant to help in the solution of the traditional problems of explanation; and second, that I believe the theory of why-questions to be basically correct as developed so far, and have rather less confidence in what follows.

Let us suppose that we are in a context with background K of accepted theory plus information, and the question Q arises here. Let Q have topic B, and contrast-class $X = \{B, C, \ldots, N\}$. How good is the answer *Because A*?

There are at least three ways in which this answer is evaluated. The first concerns the evaluation of A itself, as acceptable or as likely to be true. The second concerns the extent to which A *favors* the topic B as against the other members of the contrast-class. (This is where Hempel's criterion of giving reasons to expect, and Salmon's criterion of statistical relevance may find application.) The third concerns the comparison of *Because A* with other possible answers to the same question; and this has three aspects. The first is whether A is more probable (in view of K); the second whether it favors the topic to a greater extent; and the third, whether it is made wholly or partially irrelevant by other answers that could be given. (To this third aspect, Salmon's considerations about *screening off* apply.) Each of these three main ways of evaluation needs to be made more precise.

The first is of course the simplest: we rule out *Because A* altogether if K implies the denial of A; and otherwise ask what probability K bestows on A. Later we compare this with the probability which K bestows on the cores of other possible answers. We turn then to favoring.

If the question why B rather than $C, \ldots, N$ arises here, K must imply B and imply the falsity of $C, \ldots, N$. However, it is exactly the information that the topic is true, and the alternatives to it not true, which is irrelevant to how favorable the answer is to the topic. The evaluation uses only that part of the background information which constitutes the general theory about these phenomena, plus other "auxiliary" facts which are known but which do not imply the fact to be explained. This point is germane to all the accounts of explanation we have seen, even if it is not always emphasized. For example, in Salmon's first account, A explains B only if the probability of B given A does not equal the probability of A *simpliciter*. However, if I know that A and that B (as is often the case when I say that B because A), then my *personal probability* (that is, the probability given all the information I have) of A equals that of B and that of B given A, namely 1. Hence the probability to be used in evaluating answers is not at all the probability given all my background information, but rather, the probability given some of the general theories I accept plus some selection from my data.[8] So the evaluation of the answer *Because A* to question Q proceeds with reference only to a certain part $K(Q)$ of K. How that part is selected is equally important to all the theories of explanation I have discussed. Neither the other authors nor I can say much about it. Therefore the selection of the part $K(Q)$ of K that is to be used in the further evaluation of A, must be a further contextual factor.[9]

If $K(Q)$ plus A implies B, and implies the falsity of $C, \ldots, N$, then A receives in this context the highest marks for favoring the topic B.

Supposing that A is not thus, we must award marks on the basis of how well A redistributes the probabilities on the contrast-class so as to favor B against its alternatives. Let us call the probability in the light of $K(Q)$ alone the *prior* probability (in this context) and the probability given $K(Q)$ plus A the *posterior* probability. Then A will do best here if the posterior probability of B equals 1. If A is not thus, it may still do well provided it shifts the mass of the probability function toward B; for example, if it raises the probability of B while lowering that of $C, \ldots, N$; or if it does not lower the probability of B while lowering that of some of its closest competitors.

I will not propose a precise function to measure the extent to which the posterior probability distribution favors B against its alternatives, as compared to the prior. Two facts matter: the minimum odds of B against $C, \ldots, N$, *and* the number of alternatives in $C, \ldots, N$ to which B bears these minimum odds. The first should increase, the second decrease. Such an increased favoring of the topic against its alternatives is quite compatible with a decrease in the probability of the topic. Imagining a curve which depicts the probability distribution, you can easily see how it could be changed quite dramatically so as to single out the topic—as the tree that stands out from the wood, so to say—even though the new advantage is only a relative one. Here is a schematic example:

Why E_1 rather than E_2, ..., E_{1000}?
Because A.
$Prob\ (E_1) = \ldots = Prob\ (E_{10}) = 99/1000 = 0.099$
$Prob\ (E_{11}) = \ldots = Prob\ (E_{1000}) = 1/99,000 = 0.00001$
$Prob\ (E_1/A) = 90/1000 = 0.090$
$Prob\ (E_2/A) = \ldots = Prob\ (E_{1000}/A) = 10/999,000 = 0.00001$

Before the answer, E_1 was a good candidate, but in no way distinguished from nine others; afterwards, it is head and shoulders above all its alternatives, but has itself lower probability than it had before.

I think this will remove some of the puzzlement felt in connection with Salmon's examples of explanations that lower the probability of what is explained. In Nancy Cartwright's example of the poison ivy ("Why is this plant alive?") the answer ("It was sprayed with defoliant") was statistically relevant, but did not redistribute the probabilities so as to favor the topic. The mere fact that the probability was lowered is, however, not enough to disqualify the answer as a telling one.

There is a further way in which A can provide information which favors the topic. This has to do with what is called Simpson's Paradox; it is again Nancy Cartwright who has emphasized the importance of this for the theory of explanation.[10] Here is an example she made up to illustrate it. Let H be "Tom has heart disease"; S be "Tom smokes"; and E, "Tom does exercise." Let us suppose the probabilities to be as follows:

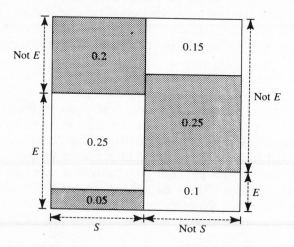

Shaded areas represent the cases in which H is true, and numbers the probabilities. By the standard calculation, the conditional probabilities are

$Prob\ (H/S) = Prob\ (H) = 1/2$
$Prob\ (H/S\ and\ E) = 1/6$
$Prob\ (H/E) = 1/8$
$Prob\ (H/S\ and\ not\ E) = 1$
$Prob\ (H/not\ E) = 3/4$

In this example, the answer "Because Tom smokes" does favor the topic that Tom has heart disease, in a straightforward (though derivative) sense. For as we would say it, the odds of heart disease increase with smoking regardless of whether he is an exerciser or a non-exerciser, and he must be one or the other.

Thus we should add to the account of what it is for A to favor B as against C, ..., N that: if $Z = \{Z_1, \ldots, Z_n\}$ is a logical partition of explanatorily relevant alternatives, and A favors B as against C, ..., N if any member of Z is added to our background information, then A does favor B as against C, ..., N.

We have now considered two sorts of evaluation: how probable is A itself? *and,* how much does A favor B as against C, ..., N? These are independent questions. In the second case, we know what aspects to consider, but do not have a precise formula that "adds them all up." Neither do we have a precise formula to weigh the importance of how likely the answer is to be true, against how favorable the information is which it provides. But I doubt the value of attempting to combine all these aspects into a single-valued measurement.

In any case, we are not finished. For there are relations among answers that go beyond the comparison of how well they do with respect to the criteria considered so far. A famous case, again related to Simpson's Paradox, goes as follows (also discussed in Cartwright's paper): at a certain university it was found that the admission rate for women was lower than that for men. Thus "Janet is a woman" appears to tell for "Janet was not admitted" as against "Janet was admitted." However, this was not a case of sexual bias. The admission rates for men and women for each department in the university were approximately the same. The appearance of bias was created because women tended to apply to departments with lower admission rates. Suppose Janet applied for admission in history; the statement "Janet applied in history" *screens off* the statement "Janet is a woman" from the topic "Janet was not admitted" (in the Reichenbach–Salmon sense of "screens off": P screens off A from B exactly if the probability of B given P and A is just the probability of B given P alone). It is clear then that the information that Janet applied in history (or whatever other department) is a much more telling answer than the earlier reply, in that it makes that reply irrelevant.

We must be careful in the application of this criterion. First, it is not important if some proposition P screens off A from B if P is not the core of an answer to the question. Thus if the why-question is a request for information about the mechanical processes leading up to the event, the answer is no worse if it is statistically screened off by other sorts of information. Consider "Why is Peter dead?" answered by "He received a heavy blow on the head" while we know already that Paul has just murdered Peter in some way. Secondly, a screened-off answer may be good but partial rather than irrelevant. (In the same example, we know that there must be some true proposition of the form "Peter received a blow on the head with impact x," but that does not disqualify the answer, it only means that some more informative answer is possible.) Finally, in the case of a deterministic process in which state A_i, and no other state, is followed by state A_{i+1}, the best answers to the question "Why is the system in state A_n at time t_n?" may all have the form "Because the system was in state A_i at time t_i," but each such answer is screened off from the event described in the topic by some other, equally good answer. The most accurate conclusion is probably no more than that if one answer is screened off by another, and not conversely, then the latter is better in some respect.

When it comes to the evaluation of answers to why-questions, therefore, the account I am able to offer is neither as complete nor as precise as one might wish. Its shortcomings, however, are shared with the other philosophical theories of explanation I know (for I have drawn shamelessly on those other theories to marshal these criteria for answers). And the traditional main problems of the theory of explanation are solved not by seeing what these criteria are, but by the general theory that explanations are answers to why-questions, which are themselves contextually determined in certain ways.

2.5. Presupposition and Relevance Elaborated

Consider the question "Why does the hydrogen atom emit photons with frequencies in the general Balmer series (only)?" This question presupposes that the hydrogen atom emits photons with these frequencies. So how can I even ask that question unless I believe that theoretical presupposition to be true? Will my account of why-questions not automatically make scientific realists of us all?

But recall that we must distinguish carefully what a theory *says* from what we believe when we accept that theory (or rely on it to predict the weather or build a bridge, for that matter). The epistemic commitment involved in accepting a scientific theory, I have argued, is not belief that it is true but only the weaker belief that it is empirically adequate. In just the same way we must distinguish what the question says (that is, *presupposes*) from what we believe when we ask that question. The example I gave above is a question which arises (as I have defined that term) in any context in which those hypotheses about hydrogen, and the atomic theory in question, are *accepted*. Now, when I ask the question, if I ask it seriously and in my own person, I imply that I believe that this question arises. But that means then only that my epistemic commitment indicated by, or involved in, the asking of this question, is exactly—no more and no less than—the epistemic commitment involved in my acceptance of these theories.

Of course, the discussants in this context, in which those theories are accepted, are conceptually immersed in the theoretical world-picture. They talk the language of the theory. The phenomenological distinction between objective or real, and not objective or unreal, is a distinction between what there is and what there is not which is drawn in that theoretical picture. Hence the questions asked are asked in the theoretical language—how could it be otherwise? But the epistemic commitment of the discussants is not to be read off from their language.

Relevance, perhaps the other main peculiarity of the why-question, raises another ticklish point, but for logical theory. Suppose, for instance, that I ask a question about a sodium sample, and my background theory includes present atomic physics. In that case the answer to the question may well be something like: because this material has such-and-such an atomic structure. Recalling this answer from one of the main examples I have used to illustrate the asymmetries of explanation, it will be noted that, *relative to* this background theory, my answer is a proposition necessarily equivalent to: because this material has such-and-such a characteristic spectrum. The reason is that the spectrum is unique—it identifies the material as having that atomic structure. But, here is the asymmetry, I could not well have answered the question by saying that this material has that characteristic spectrum.

These two propositions, one of them relevant and the other not, are equivalent relative to the theory. Hence they are true in exactly the same possible worlds allowed by the theory (less metaphysically put: true in exactly the same models of that theory). So now we have come to a place where there is a conflict with the simplifying hypothesis generally used in formal semantics, to the effect that propositions which are true in exactly the same possible worlds are identical. If one proposition is relevant and the other not, they cannot be identical.

We could avoid the conflict by saying that of course there are possible worlds which are not allowed by the background theory. This means that when we single out one proposition as relevant, in this context, and the other as not relevant and hence distinct from the first, we do so in part by thinking in terms of worlds (or models) regarded in this context as impossible.

I have no completely telling objection to this, but I am inclined to turn, in our semantics, to a different modeling of the language, and reject the simplifying hypothesis. Happily there are several sorts of models of language, not surprisingly ones that were constructed in response to other reflections on relevance, in which propositions can be individuated more finely. One particular sort of model, which provides a semantics for Anderson and Belnap's logic of tautological entailment, uses the notion of *fact*.[11] There one can say that

It is either raining or not raining

It is either snowing or not snowing

although true in exactly the same possible situations (namely, in all) are yet distinguishable through the consideration that today, for example, the first is *made true* by the fact that it is raining, and the second is made true by quite a different fact, namely, that it is not snowing. In another sort of modeling, developed by Alasdair Urquhart, this individuating function is played not by facts but by bodies of information.[12] And still further approaches, not necessarily tied to logics of the Anderson–Belnap stripe, are available.

In each case, the relevance relation among propositions will derive from a deeper relevance relation. If we use facts, for example, the relation R will derive from a request to the effect that the answer must provide a proposition which describes (is made true by) facts of a certain sort: for example, facts about atomic structure, or facts about this person's medical and physical history, or whatever.

3. Conclusion

Let us take stock. Traditionally, theories are said to bear two sorts of relation to the observable phenomena: *description* and *explanation*. Description can be more or less accurate, more or less informative; as a minimum, the facts must "be allowed by" the theory (fit some of its models), as a maximum the theory actually implies the facts in question. But in addition to a (more or less informative) description, the theory may provide an explanation. This is something "over and above" description; for example, Boyle's law describes the relationship between the pressure, temperature, and volume of a contained gas, but does not explain it—kinetic theory explains it. The conclusion was drawn, correctly I think, that even if two theories are strictly empirically equivalent they may differ in that one can be used to answer a given request for explanation while the other cannot.

Many attempts were made to account for such "explanatory power" purely in terms of those features and resources of a theory that make it informative (that is, allow it to give better descriptions). On Hempel's view, Boyle's law does explain these empirical facts about gases, but minimally. The kinetic theory is perhaps better *qua* explanation simply because it gives so much more information about the behavior of gases, relates the three quantities in question to other observable quantities, has a beautiful simplicity, unifies our overall picture of the world, and so on. The use of more sophisticated statistical relationships by Wesley Salmon and James Greeno (as well as by I. J. Good, whose theory of such concepts as weight of evidence, corroboration, explanatory power, and so on deserves more attention from philosophers), are all efforts along this line.[13] If they had succeeded, an empiricist could rest easy with the subject of explanation.

But these attempts ran into seemingly insuperable difficulties. The conviction grew that explanatory power is something quite irreducible, a special feature differing in kind from empirical adequacy and strength. An inspection of examples defeats any attempt to identify the ability to explain with any complex of those more familiar and down-to-earth virtues that are used to evaluate the theory *qua* description. Simultaneously it was argued that what science is really after is understanding, that this consists in being in a position to explain, hence what science is really after goes well beyond empirical adequacy and strength. Finally, since the theory's ability to explain provides a clear reason for accepting it, it was argued that explanatory power is evidence for the *truth* of the theory, special evidence that goes beyond any evidence we may have for the theory's empirical adequacy.

Around the turn of the century, Pierre Duhem had already tried to debunk this view of science by arguing that explanation is not an aim of science. In retrospect, he fostered that explanation-mysticism which he attacked. For he was at pains to grant that explanatory power does not consist in resources for description. He argued that only metaphysical theories explain, and that metaphysics is an enterprise foreign to science. But fifty years later, Quine having argued that there is no demarcation between science and philosophy, and the difficulties of the ametaphysical stance of the positivist-oriented philosophies having made a return to metaphysics tempting, one noticed that scientific activity does involve explanation, and Duhem's argument was deftly reversed.

Once you decide that explanation is something irreducible and special, the door is opened to elaboration by means of further concepts pertaining thereto, all equally irreducible and special. The premises of an explanation have to include lawlike statements; a statement is lawlike exactly if it implies some non-trivial counterfactual conditional statement; but it can do so only by asserting relationships of necessity in nature. Not all classes correspond to genuine properties; properties and propensities figure in explanation. Not everyone has joined this return to essentialism or neo-Aristotelian realism, but some eminent realists have publicly explored or advocated it.

Even more moderate elaborations of the concept of explanation make mysterious distinctions. Not every explanation is a scientific explanation. Well then, that irreducible explanation-relationship comes in several distinct types, one of them being scientific. A scientific explanation has a special form, and adduces only special sorts of information to explain—information about causal connections and causal processes. Of course, a causal relationship is just what 'because' must denote; and since the *summum bonum* of science is explanation, science must be attempting even to describe something beyond the observable phenomena, namely causal relationships and processes.

These last two paragraphs describe the flights of fancy that become appropriate if explanation is a relationship *sui generis* between theory and fact. But there is no direct evidence for them at all, because if you ask a scientist to explain something to you, the information he gives you is not different in kind (and does not sound or look different) from the information he gives you when you ask for a description. Similarly in "ordinary" explanations: the information I adduce to explain the rise in oil prices, is information I would have given you to a battery of requests for description of oil supplies, oil producers, and oil consumption. To call an explanation scientific, is to say nothing about its form or the sort of information adduced, but only that the explanation draws on science to get this information (at least to some extent) and, more importantly, that the criteria of evaluation of how good an explanation it is, are being applied using a scientific theory (in the manner I have tried to describe in section 2.1 above).

The discussion of explanation went wrong at the very beginning when explanation was conceived of as a relationship like description: a relation between theory and fact. Really it is a three-term relation, between theory, fact, and context. No wonder that no single relation between theory and fact ever managed to fit more than a few examples! Being an explanation is essentially relative, for an explanation is an *answer*. (In just that sense, being a daughter is something relative: every woman is a daughter, and every daughter is a woman, yet being a daughter is not the same as being a woman.) Since an explanation is an answer, it is evaluated *vis-à-vis* a question, which is a request for information. But exactly what is requested, by means of the question "Why is it the case that *P*?," differs from context to context. In addition, the background theory plus data relative to which the question is evaluated, as arising or not arising, depends on the context. And even what part of that background information is to be used to evaluate how good the answer is, *qua* answer to that question, is a contextually determined factor. So to say that a given theory can be used to explain a certain fact, is always elliptic for: there is a proposition which is a telling answer, relative to this theory, to the request for information about certain facts (those counted as relevant for *this* question) that bear on a comparison between this fact which is the case, and certain (contextually specified) alternatives which are not the case.

So scientific explanation is not (pure) science but an application of science. It is a use of science to satisfy certain of our desires; and these desires are quite specific in a specific context, but they are always desires for descriptive information. (Recall: every daughter is a woman.) The exact content of the desire, and the evaluation of how well it is satisfied, varies from context to context. It is not a single desire, the same in all cases, for a very special sort of thing, but rather, in each case, a different desire for something of a quite familiar sort.

Hence there can be no question at all of explanatory power as such (just as it would be silly to speak of the "control power" of a theory, although of course we rely on theories to gain control over nature and circumstances). Nor can there be any question of explanatory success as providing evidence for the truth of a theory that goes beyond any evidence we have for its providing an adequate description of the phenomena. For in each case, a success of explanation is a success of adequate and informative description. And while it is true that we seek for explanation, the value of this search for science is that the search for explanation is *ipso facto* a search for empirically adequate, empirically strong theories.

Notes

1. At the end of my "The Only Necessity is Verbal Necessity." *Journal of Philosophy*, 74 (1977), 71–85 (itself an application of formal pragmatics to a philosophical problem), there is a short account of the development of these ideas, and references to the literature. The paper "Demonstratives" by David Kaplan which was mentioned there as forthcoming, was completed and circulated in mimeo'd form in the spring of 1977; it is at present the most important source for the concepts and applications of formal pragmatics, although some aspects of the form in which he develops this theory are still controversial. . . .

2. Belnap's theory was first presented in *An Analysis of Questions: Preliminary Report* (Santa Monica, Cal.: System Development Corporation, technical memorandum 7–1287–1000/00, 1963), and is now more accessible in N. D. Belnap, Jr. and J. B. Steel, Jr., *The Logic of Questions and Answers* (New Haven: Yale University Press, 1976).

3. I heard the example from my former student Gerald Charlwood. Ian Hacking and J. J. C. Smart told me that the officer was Sir Charles Napier.

4. C. L. Hamblin. "Questions," *Australasian Journal of Philosophy*, 36 (1958), 159–68.

5. The defining clause is equivalent to "any proposition which is true if any direct answer to Q is true." This includes, of course, propositions which would normally be expressed by means of 'metalinguistic' sentences—a distinction which, being language-relative, is unimportant.

6. In the book by Belnap and Steel (see n. 2 above), Bromberger's theory of why-questions is cast in the general form common to elementary questions. I think that Bromberger arrived at his concept of 'abnormic law' (and the form of answer exhibited by '"Grünbaum" is spelled with an umlaut because it is an English word borrowed from German, and no English words are spelled with an umlaut except those borrowed from another language in which they are so spelled'), because he ignored the tacit *rather than* (contrast-class) in why-interrogatives, and then had to make up for this deficiency in the account of what the answers are like.

7. I call this a matter of regimentation, because the theory could clearly be developed differently at this point, by building the claim of relevance into the answer as an explicit conjunct. The result would be an alternative theory of why-questions which, I think, woud equally save the phenomena of explanation or why-question asking and answering.

8. I mention Salmon because he does explicitly discuss this problem, which he calls *the problem of the reference class*. For him this is linked with the (frequency) interpretation of probability. But it is a much more general problem. In deterministic, non-statistical (what Hempel called a deductive-nomological) explanation, the adduced information implies the fact explained. This implication is relative to our background assumptions, or else those assumptions are part of the adduced information. But clearly, our information that the fact to be explained is actually the case, and all its consequences, must carefully be kept out of those background assumptions if the account of explanation is not to be trivialized. *Mutatis mutandis* for statistical explanations given by a Bayesian, as is pointed out by Glymour in his *Theory and Evidence*.

9. I chose the notation $K(Q)$ deliberately to indicate the connection with models of

rational belief, conditionals, and hypothetical reasoning, as discussed for example by William Harper. There is, for example, something called the Ramsey test: to see whether a person with total beliefs K accepts that if A then B, he must check whether $K(A)$ implies B, where $K(A)$ is the 'least revision' of K that implies A. In order to 'open the question' for A, such a person must similarly shift his beliefs from K to $K?A$, the 'least revision' of K that is consistent with A; and we may conjecture that $K(A)$ is the same as $(K?A)\&A$. What I have called $K(Q)$ would, in a similar vein, be a revision of K that is compatible with every member of the contrast class of Q, and also with the denial of the topic of Q. I don't know whether the 'least revision' picture is the right one, but these suggestive similarities may point to important connections; it does seem, surely, that explanation involves hypothetical reasoning. Cf. W. Harper, "Ramsey Test Conditionals and Iterated Belief Change," pp. 117–35, in W. Harper and C. A. Hooker, *Foundations of Probability Theory, Statistical Inference, and Statistical Theories of Science* (Dordrecht: Reidel, 1976), and his "Rational Conceptual Change," in F. Suppe and P. Asquith (eds.), *PSA 1976* (East Lansing: Philosophy of Science Association, 1977).

10. Nancy Cartwright, "Causal Laws and Effective Strategies," *Nous,* 8 (1979), 419–37.

11. See my "Facts and Tautological Entailment," *Journal of Philosophy,* 66 (1969), 477–87 and reprinted in A. R. Anderson and N. D. Belnap, Jr., *Entailment* (Princeton: Princeton University Press, 1975), and "Extension, Intension, and Comprehension," M. Munitz (ed.), *Logic and Ontology* (New York: New York University Press, 1973).

12. For this and other approaches to the semantics of relevance see Anderson and Belnap, op. cit. (n. 11 above).

13. I. J. Good, "Weight of Evidence, Corroboration, Explanatory Power, and the Utility of Experiments," *Journal of the Royal Statistical Society,* series B, 22 (1960), 319–31; and "A Causal Calculus," *British Journal for the Philosophy of Science,* 11 (1960/61), 305–18 and 12 (1961/62), 43–51. For discussion, see W. Salmon, "Probabilistic Causality," *Pacific Philosophical Quarterly,* 1980.

7

Theoretical Explanation

Wilfrid Sellars

I

My purpose in this paper is to discuss some of the features of theoretical explanation which have been at the center of the philosophical stage in recent years. Since the term 'theory' covers importantly different types of explanatory systems, each of which is capable of generating philosophical perplexities, let me make it clear from the outset that I shall limit myself to theories of the type which, to speak informally, explain the behavior of objects of a certain domain by "identifying" these objects with systems of objects of another domain, and deriving the laws governing the objects of the first domain from the fundamental laws governing the objects of the second domain. This schema, which has been deliberately designed to highlight the topics which I wish to discuss, covers two importantly different but, as I hope to show, importantly similar, types of theoretical explanation:

1. The explanation of the behavior of "observable" things in terms of "unobservable" things, as in the kinetic theory of gases.
2. The explanation of the behavior of the "unobservable" things of one theoretical framework in terms of the "unobservable" things of another theoretical framework, as in the explanation of the behavior of chemical substances in terms of atomic physics.

I have put the expressions "observable" and "unobservable" in quotation marks because they are technical terms which differ in important respects from their counterparts in ordinary usage. Thus it is not absurd to speak of observing viruses and protein molecules through an appropriately constructed electron microscope. But, as is evident, this extended use of the term is built on the physical theory of the instrument and how it relates to the physical systems which can be observed by its use. Again, to identify the objects observed by its use as "protein molecules" or "viruses" presupposes biochemical theory and pathology. Furthermore, it is particularly clear that the observations made by the use of an instrument cannot be the grounds on which we accept the theory of the instrument. For until we have the

Reproduced from *Philosophy of Science: The Delaware Seminar 1963, Vol. 2* by permission of the University of Delaware.

theory of the instrument, we *logically* can't make observations with it, though, of course, we can observe it and its behavior as a perceptible physical object.

Thus, although observation in the extended sense provides data for the elaboration of theories pertaining to objects which are not observable in the absence of theory-laden instrumentation, the concept of such observation presupposes the concept of unaided perception, and it is the latter, as submitted to scientific discipline, which the philosopher of science refers to as observation in his technical sense. There are many problems pertaining to unaided sense perception which are still matters of dispute among philosophers. I shall avoid these problems and simply assume what practicing scientists take for granted, that in sense perception we have direct access to public physical objects and processes, and can distinguish between favorable and unfavorable circumstances for telling *by looking, listening,* etc., what perceptible characteristics can reliably be ascribed to these objects and processes. It is customary to call these perceptible characteristics "observable properties" and their linguistic expression "observation predicates." The concept of what one can tell about an object or process by looking, listening, etc., does not cover the same ground in all contexts. Thus a clinical psychologist can tell by looking that a patient is depressed, and many people can tell by looking that a certain object is 10 feet high. Reliability and intersubjectivity with respect to a given science at a given stage of its development determine what, for that science, can be told by looking. Implicit in the above discussion are distinctions which can be tabulated as follows:

1. Observable things and processes.
2. The observable properties of observable things and processes.
3. Unobservable things and processes.
4. The unobservable properties of unobservable things and processes.

This classification makes no pretense of being exhaustive. Indeed, a glance at it cannot help but raise the questions:

1. Can observable things and processes have unobservable properties?
2. Can unobservable things and processes have observable properties?

The second question is by no means unambiguous, but a possible example would be

Can a system of subatomic particles be liquid?

I shall say something about questions like this at the end of the [chapter]. Of more immediate interest is the first question. The answer is clearly affirmative. There are, however, two types of unobservable property which an observable object might have

1. Specific gravity.
2. The property of consisting of subatomic particles.

The second type is problematic. In effect, to say that observable objects can have such a property implies an affirmative answer to the question, "Can a system of subatomic particles have an observable property?" We shall accordingly postpone considering this type of unobservable property of observable objects until we take up the latter question. It is worth pointing out, however, that if our initial characterization of one of the varieties of theory with which we shall be concerned (i.e., theories which explain the behavior of a domain of observable objects by *identifying* them with systems of certain unobservable objects) is taken at its face value, it must be true that

> 1. Observable objects have unobservable properties of the second type listed above.
> 2. Systems of unobservable objects have observable properties.

What this does is bring out the full force of the word "identity" and, as a result, generates some measure of perplexity with the idea that to accept chemical theory is to accept the idea that the water in a glass of water is identical with a quantity of H_2O.

II

But before we can make any headway in this and other issues pertaining to theories, additional distinctions must be drawn. The best place to begin is with the first category of unobservable characteristics which observable objects can have. The ones I have in mind have in common the fact that they are explicitly definable in terms of observable properties. Here the important thing to note is the revolution which modern logic has brought to the concept of explicit definition. The days when the paradigm of a definition was

x is a man $=_{df.}$ x is a living organism *and* x is rational

have gone forever. Notice that the right-hand side of this definition includes the logical expression "and." The revolution in the theory of definition brought with it, among other things, a recognition of the importance of definitions in which other logical expressions play the role played above by "and." Modern philosophy of science lays particular stress on definitions in which the right-hand side is conditional in form, thus:

x is water soluble $=_{df.}$ if x is put in water, then x dissolves.

So-called *operational definitions* have this form. It is sometimes thought that the use of operationally-defined terms introduces an anthropomorphic element into the content of science by introducing a reference to the activities of scientists into the very concepts in terms of which the world is to be scientifically described. This is a simple misunderstanding. Compare the two definitions

> (i) x is water soluble $=_{df.}$ if x is put in water, then x dissolves
> (ii) x is water soluble $=_{df.}$ if x comes to be in water, then x dissolves

The first does contain an anthropomorphic element, for *putting things in water* is something people *do*. On the other hand, the second definition, though conditional in form, has in place of "is put," "comes to be," and has no anthropomorphic content. To every operational definition there corresponds a conditional definition which makes no reference to operations. Thus, operational definitions are that subspecies of conditional definitions where the antecedent is a state of affairs which can be realized by a doing on the part of the scientist. It is, therefore, convenient to formulate them in terms of this doing. This *methodologically* important formulation, however, can always be discounted and the operation word replaced by an expression making no reference to human activity.

It is sometimes thought that even observation predicates should be said to be "operationally defined" because their importance lies in the fact that one can tell that they apply to something by "performing the operation of looking (in controlled circumstances)." The reason given is a true proposition, but it does not support the proposal, which blurs the distinction between predicates which, whether or not they are defined in terms of operations, can be determined to apply to an object by performing an operation. These two notions do not coincide. The latter is more inclusive than the former. And if one takes "pencil and paper operations" into account, the class of predicates the applicability of which to an object can be determined by performing an operation becomes unilluminatingly large.

Let us introduce the expression "empirical predicate" to cover not only observation predicates but predicates which are directly or indirectly defined in terms of observation predicates and, of course, the vocabulary of logic. An empirical property is one the linguistic expression for which is an empirical predicate. Defined empirical predicates are a subset of empirical predicates, and operationally defined empirical predicates a subset of defined empirical predicates.

Fruitful empirical predicates are those which occur in inductively confirmed generalizations. As Professor Bergmann has emphasized, any predicate which is defined in terms of well-formed expressions consisting of observation predicates and logical words is meaningful in the sense that it stands for a property. On the other hand, it may be meaningless in the sense of pointless or lacking scientific significance.

The theory of measurement is a large topic about which it is better to say nothing rather than little. I must, however, call attention to the fact that what might be called ground floor metrical predicates are operationally defined. Yet, as Hempel and others have pointed out, these operationally defined metrical predicates are tied in ways which have not yet been fully analyzed, to metrical predicates which involve the real number system. If one compares, as seems reasonable, the relation of these "idealized" metrical predicates to operationally defined metrical predicates to that of theoretical to empirical predicates generally, then already in the theory of measurement one runs into the problem of the reality of theoretical constructs. It will be clear from the general tenor of this paper that in this context also I would be strongly inclined to give an affirmative answer. But there are sticky issues in this neighborhood, and I am happy that I can sweep this issue under the rug.

III

There are many interesting questions relating to operationally defined predicates. Some of these are logical; thus it is pointed out that to say that something is soluble is to say that if it *were* put in water it *would* dissolve, which confronts one with the problem of subjunctive conditionals and causal modalities. Then there is the problem of how one establishes that a particular object has a conditional property such as is defined by an operational definition. If one has an inductive generalization to the effect that all objects of kind K are soluble, and knows that this object is of kind K, then one is indeed entitled to say that this object is soluble. But this pushes the problem one stage back, for, it would seem, the evidence for the generalization that all K's are soluble must include statements of the form "this K is soluble" which are known to be true independently of the generalization. It must, it would seem, be possible to know about a particular object that it has a certain conditional (or, as it usually called, dispositional) property without reference to any kind to which it may or may not belong. Thus, if we perform the relevant operation repeatedly on the object and as often get the appropriate result, we can argue inductively,

At t_1 this was O'd and R ensued

At t_2 it was O'd and R ensued

· · · · · · · · · · · · · · · · · ·

At t_n it was O'd and R ensued

Therefore (supposing this to be all the evidence) it is probable
that if it were *now* O'd, then R would ensue

This induction, if repeated successfully with other objects which were all of a certain kind to which the original object belongs, would authorize the generalization

If things of kind K are O'd, then R ensues

that is,

Things of kind K have the property D

where D is operationally defined in terms of O and R.

This line of thought works reasonably well with conditional properties like elasticity, but runs into trouble with "inflammable." Here the induction must rather be of the form

This piece of paper was heated and burned

That piece of paper was heated and burned

· ·

The other piece of paper was heated and burned

Therefore, probably pieces of paper are inflammable.

Here the kind, the operation, and the result are all involved in the basic induction. Needless to say, the structure of such an argument, to be plausible, would be far more complicated, but these hints should help relieve any perplexities of principle.

IV

Let us now focus our attention on the idea of a set of inductively confirmed generalizations pertaining to a certain domain of objects. These generalizations may be either "universal" or "statistical," but since the topics I shall discuss, at least in the scale in which I shall treat them, do not hinge on this distinction, I shall assume that they are universal in form, and represent them as follows:

$$G_1. \text{ if } A \text{ then } B$$
$$G_2. \text{ if } C \text{ then } D$$
$$\cdots \cdots \cdots \cdots$$
$$G_n. \text{ if } X \text{ then } Y$$

where the uppercase letters in the if-then statements represent empirical predicates, and "if A then B" is to be read: If any object or group of objects of the domain has the property A, then the object or group of objects has the property B. "A" and "B" may, of course, stand for empirical properties of any degree of complexity. Spatial and temporal relations are tacitly included.

Now according to the standard account of theories of the kind we are considering, a theoretical framework which is to explain generalizations G_1 to G_n will consist of two sets of propositions:

1. A deductive system consisting of postulates and the theorems they logically imply. The vocabulary of the system consists of two parts: (a) the vocabulary of logic and mathematics, which has its ordinary sense, and (b) the distinctive vocabulary of the theory, which, in its turn consists of two parts: (b_1) primitive terms, (b_2) defined terms. Taken simply as belonging to the deductive system, both primitive and defined terms are *uninterpreted*. This does not mean—*pace* Nagel—that they are *variables*.

2. A set of correspondence rules which correlate some expressions belonging to (b_2), above, with empirical predicates which occur in $G_1 \ldots G_n$. These rules do not suffice to correlate the primitive vocabulary of the theory with empirical predicates. If they did, the "theory," if successful, would simply be a representation of empirical generalizations in the form of a deductive system.

The correspondence rules are so chosen that ideally they correlate the inductively established generalizations $G_1 \ldots G_n$ with theorems in the deductive system, and correlate no theorem in the deductive system with an inductively disconfirmed empirical generalization. If it succeeds in doing this, the theory is insofar confirmed. If, moreover, the correspondence rules correlate a theorem in the deductive system with a lawlike statement in the empirical framework which has not yet been put to the test, and it is then put to the test and is inductively confirmed, this is taken to be a particularly striking confirmation of the theory.

Now the deductive system of a theory is often formulated with reference to a model. A domain of objects is pointed to, which either behaves in ways which satisfy the postulates of the deductive system, or can be imagined to do so without an absurdity which would deprive the reference to them of any value or point. The model serves a number of purposes. The most obvious is to make the theory intuitive and aid the imagination in working with it. But more than this it fills an important need in that whereas the basic magnitudes of the empirical framework are operationally defined and are therefore rooted in a background of qualitative content, the basic magnitudes of the theoretical framework, in the absence of a model, would in no way point to a foundation in nonmetrical, qualitative distinctions which might stand

to them as the qualitative dimensions of observable things stand to the metrical properties which are operationally defined with respect to them. The theory would leave
them "abstract" in a sense which reminds us of Whitehead's charge of "vacuous
actuality" against scientific materialism. The basic magnitudes of the theory would
simply point forward to the more complex theoretical properties which can be defined
in terms of them, and would find their *be all and end all* in the theorems which save
the appearances (empirical generalizations). Now by virtue of their visualizable character, riodels provide a surrogate for the "qualitative" predicates which must, in the
last analysis, be the underpinning of theoretical magnitudes if they are to be the sort
of thing that could "really exist," if this phrase can be given a stronger interpretation
than that of the irenic instrumentalist. Needless to say, the qualitative dimensions
which provide the content for the metrical form of theoretical entities need not be
the perceptual qualities of the model. It would be odd if the only qualitative dimensions of the world were those which are, in the last analysis, tied to the sensory
centers of the human brain.

Thus, however important the heuristic function of the model, and however important the layer of analogical meaning which the theoretical predicates acquire by
being explained in terms of the model, the theory is not about the objects of the
model, nor do theoretical predicates stand for properties of the objects in the model.
The reference of the theory, if it can be said to have reference, and the meanings
of the predicates of the theory, insofar as these are more than an adumbration of
things to come, are to be understood in terms of the deductive system and the co-
ordination of the theory with the empirical generalizations it is designed to explain.
This brings me to the topic of correspondence rules.

V

The first thing to note is that our previous characterization of correspondence rules
was too narrow. We need a more inclusive concept of which the type of rule referred
to above is a special case. Correspondence rules in this broader sense are rules cor-
relating theoretical predicates with empirical predicates. Within this broader classification we can distinguish:

1. Rules which correlate predicates in the theory with empirical predicates pertaining to the objects of the domain for which the theory is a theory, where the
important thing about these empirical predicates is that they occur in empirical laws
pertaining to these objects. It is not important that these predicates be *observation*
predicates or related to observation predicates by particularly short definitional chains,
nor that these definitions be operational. I shall call these rules *substantive correspondence rules*.

2. Rules which correlate predicates in the theory with predicates which, though
empirical, need not pertain to the domain of objects for which the theory is a theory.
(They may pertain, for example, to an instrument, e.g., a spectroscope.) In the case
of rules of this type, it is essential that the empirical predicates be observation predicates or related to them by operational definitions. I shall call these rules *methodological correspondence rules*.

The distinction stands out more clearly in terms of illustrations. Thus a correspondence rule of the substantive type would be

Temperature of gas   <————>   Mean kinetic energy of
in region *R* is   molecules in region
such and such   *R* is such and such

where, although it is true that since the temperature of a particular gas can be cal-
culated from experimental data, the correspondence rule provides a path which can
take us from an empirical description to a theoretical description, the primary point
of the rule is not of this methodological character, but rather to permit a correlation
of a theorem in the deductive system with an empirical law, thus the Boyle-Charles
law.

Compare, on the other hand,

Spectroscope appropriately   <————>   Atoms in region *R*
related to gas shows   are in such and such
such and such lines   a state of excitation

Correspondence rules of this kind differ from the above in that whereas in the
former case it would not be absurd to say that the rule in some sense *identifies*
temperature with the mean kinetic energy of its molecular constituents, it would be
absurd to say of the latter that it *identifies* spectral lines with the state of excitation
of the atoms. Again, since spectroscopes are not gases, they do not belong to the
domain of which the theory is a theory. On the other hand, the spectroscope must
be related to a gas in an appropriate manner, and we can certainly say that the
property of causing certain spectral lines, which *is* a property, albeit relational, of
an object belonging to the domain of the theory, is correlated by the correspondence
rule with the state of excitation of the atoms. But all this does is make manifest that
the theory of the instrument enables a detection of the theoretical state by virtue of
a connection of that state with processes (electromagnetic vibrations) which can be
registered by the instrument.

It is important to note that whereas in the example used, the connection between
the observable spectrum and the theoretical state of the gas involves a well-developed
and articulated theory, this need not be the case. Correspondence rules of the methodological type can be built on very schematic and promissory-noteish conception
of *how* the observable phenomena are connected with the theoretical states.

It is correspondence rules of the first or substantive kind which have fascinated
philosophers of science. As I pointed out above, it is not implausible to formulate
them as statements of identity. Yet here we must be careful, for while it is *very*
plausible to say that gases *are* populations of molecules, it is by no means as plausible to make the same move with respect to empirical and theoretical *properties*.
However tempting the idea might appear to start with, it is difficult to see just what
could be meant by saying that a property defined in terms of observable characteristics is identical with a property defined in terms of theoretical primitives. Could
one, perhaps, take seriously the idea that gases are identical with populations of
molecules, while denying that the empirical properties of gases are identical with
the theoretical properties of populations of molecules? This move is sometimes made.
But if, as I suspect, the two identifications are inseparable, the *prima facie* implau-
sibility of the identification of empirical and theoretical properties must militate against
the *prima facie* plausibility of the identification of empirical and theoretical objects.

VI

I think that some light can be thrown on this puzzle by considering the case where it is not a question of "identifying" empirical properties with the corresponding theoretical properties, but rather properties defined in one theoretical framework with properties defined in a second theoretical framework to which the first theory is said to be "reducible." The stock example is the reducibility of the objects of current chemical theory to complexes of the objects of current atomic physics. To speak of this reduction as accomplished fact is probably to idealize the actual situation, but since philosophers are always concerned with what might in principle be the case, to understand what the structure of such a reduction would be if it were to exist will serve our purpose.

In speaking of reduction, one must be very careful not to think of the reduction of the content of one scientific theory to that of another as amounting to the reduction of one *science* to another. For even if the former reduction been successfully achieved, the two sciences will normally remain distinct at the empirical level. Thus the science whose theory has been reduced will concern itself as before with a narrower domain of empirical objects (chemical substances and processes), will use different experimental techniques, and will gain access to concepts in the unified theory by different operational routes and at a different level of the theory.

But before spelling out these differences in more detail, let us see how the theme of *identification,* which is present in the concept of *reduction,* is to be understood. If we suppose that the two theories prior to the reduction have been formulated as deductive systems, then the distinctive vocabulary of each theory had no definitional connection with that of the other. The only connection between them was "round about," in that the domain of empirical objects with which the one theory (chemistry) was concerned is a subset of the domain with which the other (atomic physics) is (in principle) concerned.

But if, as the concept of reduction implies, theoretical propositions in chemistry are to be derived from theoretical propositions in atomic physics, this unconnectedness of the two vocabularies must be overcome. Chemical expressions must be correlated with physical expressions. Thus, if a certain chemical process is to be explained in terms of atomic physics, the statement formulating the law governing this process must be correlated with a sentence derivable in atomic theory which is to be the explanation of this process. This correlation will consist of rules which are like the substantive correspondence rules we have been considering, with the difference that they do not correlate theoretical with empirical predicates, but expressions in one theory with expressions in another. Again, like the correspondence rules we have been considering, they are not logically true; they are not true by virtue of the meaning of the expressions they correlate. And also like them they are adopted on scientific and, therefore, broadly speaking, empirical grounds. This latter fact is ultimately Nagel's ground for calling them "bridge laws," which I find a most misleading expression. I propose, instead, to introduce a still broader sense of "correspondence rule" in accordance with which what I have been calling substantive correspondence rules can connect either a theoretical predicate with an empirical predicate (which is the case we have already considered) or a predicate in one theory with a predicate in another (which is the case we are now considering).

Now although the correspondence rules which connect theory with theory are not true by virtue of the *antecedent* meanings of the two sets of theoretical expressions, the situation involves a dimension of linguistic free play which is not present, at least at first sight, in the case of rules connecting theoretical with empirical expressions. For as far as the deductive cores of the two theories is concerned, there is no reason why the two theoretical vocabularies built from two sets of theoretical primitives could not be *replaced* by one vocabulary built from one set of primitives—those of atomic physics. The primitive vocabulary of unreduced chemical theory would reappear; this time, however, not as primitives but as defined terms in the reducing theory. The inter-theory correspondence rules would have been replaced by definitions. Notice that before this step was taken, we would be in a frame of mind similar to that in which we found ourselves with respect to kinetic theory. On the one hand, we would be strongly inclined to say that the *objects* of chemical theory are identical with systems of objects of the atomic physical theory. (Thus compare "a minimum quantity of hydrogen is an atomic system consisting of such and such a nucleus and one electron" with "a gas is a population of molecules.") On the other hand, we would clearly *not* be entitled to say that the predicates of chemical theory have the *same sense* or stand for the *same properties* as any predicates of atomic physics.

But in the case of the reduction of one theory to another we can *bring it about* that the identity statements are unproblematic, both with respect to objects and with respect to properties, by making one unified vocabulary do the work of two. We give the distinctive vocabulary of chemistry a new use in which the physical and the chemical expressions which occurred in substantive correspondence rules now have the same sense.

Yet in explicating this identity of sense we must not forget what was said above about the difference between reducing a theory and reducing a science. It was pointed out that to reduce chemical theory to physical theory is by no means to reduce chemistry to physics. The relacement of the substantive correspondence rules which related the original theories by explicit definitions leaves relatively untouched the correspondence rules, both substantive and methodological which tied the two theories to their empirical domains. To use a picture, a situation in which two balloons (theories), one above the other, each tied to the ground at a different set of places (theoretical-empirical correspondence rules), and each vertically tied to the other by various ropes (theoretical-theoretical correspondence rules), has been replaced by a situation in which there is one taller balloon which, however, is tied to the ground by two sets of ropes, one of which connects the lower part of the balloon to the places where the lower balloon was tied, while the other connects the higher part of the balloon to the places where the upper balloon was tied.

VII

If we put this by saying that the language of identity is appropriate to an informal formulation of substantive correspondence rules of the theory-theory type because they are candidates for ultimate replacement by definitional identities, the question arises, "Are the substantive correspondence rules which relate empirical predicates pertaining to gases to theoretical predicates pertaining to populations of molecules capable of similar treatment?" This, as I see it, is the heart of the problem of the reality of theoretical entities. It is, in effect, the problem "Can observable things be reduced (in principle) to the objects of physical theory?" Notice that if the substantive correspondence rules of a physical theory adequate to such a reduction were to be replaced by definitions, it would be a matter of defining the empirical vocabulary in terms of the theoretical vocabulary, and not vice versa. The impossibility of defining theoretical concepts in observational terms has been a cornerstone of the argument.

Does it make sense to speak of turning empirical predicates—and in particular observation predicates—into definitional abbreviations of complex theoretical locutions? Could observation predicates be so treated while continuing to play their perceptual role as conditioned responses to the environment? I see no reason in principle why this should not be the case.

From the standpoint of the methodology of developing science, it might seem foolish to build physical theory into the language of observation and experiment. A tentative correlation of theoretical and empirical terms would seem more appropriate than redefinition. But this is a truism which simply explains what we mean by developing science. But the perspective of the philosopher cannot be limited to that which is methodologically wise for developing science. He must also attempt to envisage the world as pictured from that point of view—one hesitates to call it Completed Science—which is the regulative ideal of the scientific enterprise. As I see it, then, substantive correspondence rules are anticipations of definitions which it would be inappropriate to implement in developing science, but the implementation of which in an ideal state of scientific knowledge would be the achieving of a unified vision of the world in which the methodologically important dualism of observation and theoretical frameworks would be transcended, and the world of theory and the world of observation would be one.

In my opinion, also, the only alternative to this conception is the instrumentalist conception of theories as deductive systems, the distinctive vocabulary of which consists of what, in the context of pure geometry, are called uninterpreted expressions, doomed as a matter of principle to remain so. As functioning in the theory they would have a *use,* but this use would be simply that of serving as essential cogs in a syntactical machine which provides an external systematization of empirical statements. The instrumentalist conception of theories correctly stresses that theories have this use and are *meaningful* in the sense that they have this use. But surely our willingness to use the language of identity in connection with empirical and theoretical objects involves a commitment which goes beyond anything which would be implied by correspondence rules if these were formulated ascetically, in accordance with instrumentalist convictions, as syntactical bridges between a *language* and a *calculus.* I do not think that this willingness rests on a mistake.[1]

Note

1. For an application of this approach to the mind-body problem, see Chapter XV of my *Science, Perception and Reality*. London: Routledge and Kegan Paul, 1968. For a discussion of explanation in terms of thing kinds and causal properties, see also Counterfactuals, dispositions and the causal modalities, in *Concepts, Theories and the Mind-Body Problem,* ed., Herbert Feigl, Grover Maxwell, and Michael Scriven, Minneapolis, 1958, especially Part II, sections 46–54.

8

Explanatory Unification

Philip Kitcher

1. The Decline and Fall of the Covering Law Model

One of the great apparent triumphs of logical empiricism was its official theory of explanation. In a series of lucid studies (Hempel 1965, Chapters 9, 10, 12; Hempel 1962; Hempel 1966), C. G. Hempel showed how to articulate precisely an idea which had received a hazy formulation from traditional empiricists such as Hume and Mill. The picture of explanation which Hempel presented, the *covering law model,* begins with the idea that explanation is derivation. When a scientist explains a phenomenon, he derives (deductively or inductively) a sentence describing that phenomenon (the *explanandum* sentence) from a set of sentences (the *explanans*) which must contain at least one general law.

Today the model has fallen on hard times. Yet it was never the empiricists' whole story about explanation. Behind the official model stood an unofficial model, a view of explanation which was not treated precisely, but which sometimes emerged in discussions of theoretical explanation. In contrasting scientific explanation with the idea of reducing unfamiliar phenomena to familiar phenomena, Hempel suggests this unofficial view: "What scientific explanation, especially theoretical explanation, aims at is not [an] intuitive and highly subjective kind of understanding, but an objective kind of insight that is achieved by a systematic unification, by exhibiting the phenomena as manifestations of common, underlying structures and processes that conform to specific, testable, basic principles" (Hempel 1966, p. 83; see also Hempel 1965, pp. 345, 444). Herbert Feigl makes a similar point: "The aim of scientific explanation throughout the ages has been *unification,* that is, the comprehending of a maximum of facts and regularities in terms of a minimum of theoretical concepts and assumptions" (Feigl 1970, p. 12).

This unofficial view, which regards explanation as unification, is, I think, more promising than the official view. My aim in this paper is to develop the view and to present its virtues. Since the picture of explanation which results is rather complex, my exposition will be programmatic, but I shall try to show that the unofficial view can avoid some prominent shortcomings of the covering law model.

Why should we want an account of scientific explanation? Two reasons present

Reprinted from *Philosophy of Science,* 48, (1981) pp. 507–531. Copyright © 1981 by the Philosophy of Science Association. With permission of the Philosophy of Science Association and the author.

themselves. Firstly, we would like to understand and to evaluate the popular claim that the natural sciences do not merely pile up unrelated items of knowledge of more or less practical significance, but that they increase our understanding of the world. A theory of explanation should show us *how* scientific explanation advances our understanding. (Michael Friedman cogently presents this demand in his (1974)). Secondly, an account of explanation ought to enable us to comprehend and to arbitrate disputes in past and present science. Embryonic theories are often defended by appeal to their explanatory power. A theory of explanation should enable us to judge the adequacy of the defense.

The covering law model satisfies neither of these *desiderata*. Its difficulties stem from the fact that, when it is viewed as providing a set of necessary *and sufficient* conditions for explanation, it is far too liberal. Many derivations which are intuitively non-explanatory meet the conditions of the model. Unable to make relatively gross distinctions, the model is quite powerless to adjudicate the more subtle considerations about explanatory adequacy which are the focus of scientific debate. Moreover, our ability to derive a description of a phenomenon from a set of premises *containing a law* seems quite tangential to our understanding of the phenomenon. Why should it be that exactly those derivations which employ laws advance our understanding?

The unofficial theory appears to do better. As Friedman points out, we can easily connect the notion of unification with that of understanding. (However, as I have argued in my (1976), Friedman's analysis of unification is faulty; the account of unification offered below is indirectly defended by my diagnosis of the problems for his approach.) Furthermore, as we shall see below, the acceptance of some major programs of scientific research—such as, the Newtonian program of eighteenth-century physics and chemistry, and the Darwinian program of nineteenth-century biology—depended on recognizing promises for unifying, and thereby explaining, the phenomena. Reasonable skepticism may protest at this point that the attractions of the unofficial view stem from its unclarity. Let us see.

2. Explanation: Some Pragmatic Issues

Our first task is to formulate the problem of scientific explanation clearly, filtering out a host of issues which need not concern us here. The most obvious way in which to categorize explanation is to view it as an activity. In this activity we answer the actual or anticipated questions of an actual or anticipated audience. We do so by presenting reasons. We draw on the beliefs we hold, frequently using or adapting arguments furnished to us by the sciences.

Recognizing the connection between explanations and arguments, proponents of the covering law model (and other writers on explanation) have identified explanations as special types of arguments. But although I shall follow the covering law model in employing the notion of argument to characterize that of explanation, I shall not adopt the ontological thesis that explanations are arguments. Following Peter Achinstein's thorough discussion of ontological issues concerning explanation in his (1977), I shall suppose that an explanation is an ordered pair consisting of a proposition and an act type.[1] The relevance of arguments to explanation resides in the fact that what makes an ordered pair (*p*, explaining *q*) an explanation is that a sentence expressing *p* bears an appropriate relation to a particular argument. (Achinstein shows how the central idea of the covering law model can be viewed in this way.) So I am supposing that there are acts of explanation which draw on arguments supplied by science, reformulating the traditional problem of explanation as the question: What features should a scientific argument have if it is to serve as the basis for an act of explanation?[2]

The complex relation between scientific explanation and scientific argument may be illuminated by a simple example. Imagine a mythical Galileo confronted by a mythical fusilier who wants to know why his gun attains maximum range when it is mounted on a flat plain, if the barrel is elevated at 45° to the horizontal. Galileo reformulates this question as the question of why an ideal projectile, projected with fixed velocity from a perfectly smooth horizontal plane and subject only to gravitational acceleration, attains maximum range when the angle of elevation of the projection is 45°. He defends this reformulation by arguing that the effects of air resistance in the case of the actual projectile, the cannonball, are insignificant, and that the curvature of the earth and the unevenness of the ground can be neglected. He then selects a kinematical argument which shows that, for fixed velocity, an ideal projectile attains maximum range when the angle of elevation is 45°. He adapts this argument by explaining to the fusilier some unfamiliar terms ("uniform acceleration," let us say), motivating some problematic principles (such as the law of composition of velocities), and by omitting some obvious computational steps. Both Galileo and the fusilier depart satisfied.

The most general problem of scientific explanation is to determine the conditions which must be met if science is to be used in answering an explanation-seeking question Q. I shall restrict my attention to explanation-seeking why-questions, and I shall attempt to determine the conditions under which an argument whose conclusion is S can be used to answer the question "Why is it the case that S?" More colloquially, my project will be that of deciding when an argument explains why its conclusion is true.[3]

We leave on one side a number of interesting, and difficult issues. So, for example, I shall not discuss the general relation between explanation-seeking questions and the arguments which can be used to answer them, nor the pragmatic conditions governing the idealization of questions and the adaptation of scientific arguments to the needs of the audience. (For illuminating discussions of some of these issues, see Bromberger 1962.) Given that so much is dismissed, does anything remain?

In a provocative article, (van Fraassen 1977) Bas van Fraassen denies, in effect, that there are any issues about scientific explanation other than the pragmatic questions I have just banished. After a survey of attempts to provide a theory of explanation he appears to conclude that the idea that explanatory power is a special virtue of theories is a myth. We accept scientific theories on the basis of their empirical adequacy and simplicity, and, having done so, we use the arguments with which they supply us to give explanations. This activity of applying scientific arguments in explanation accords with extra-scientific, "pragmatic," conditions. Moreover, our views about these extra-scientific factors are revised in the light of our acceptance of new theories: ". . . science schools our imagination so as to revise just those prior judgments of what satisfies and eliminates wonder" (van Fraassen 1977, p. 150). Thus there are no context-independent conditions, beyond those of simplicity and empirical adequacy which distinguish arguments for use in explanation.

Van Fraassen's approach does not fit well with some examples from the history of science—such as the acceptance of Newtonian theory of matter and Darwin's theory of evolution—examples in which the explanatory promise of a theory was appreciated in advance of the articulation of a theory with predictive power. (See pp. 170–172.) Moreover, the account I shall offer provides an answer to skepticism that no "global constraints" (van Fraassen 1977, p. 146) on explanation can avoid the familiar problems of asymmetry and irrelevance, problems which bedevil the covering law model. I shall try to respond to van Fraassen's challenge by showing that there are certain context-independent features of arguments which distinguish them for application in response to explanation-seeking why-questions, and that we can assess theories (including embryonic theories) by their ability to provide us with such arguments. Hence I think that it is possible to defend the thesis that historical appeals to the explanatory power of theories involve recognition of a virtue over and beyond considerations of simplicity and predictive power.

Resuming our main theme, we can use the example of Galileo and the fusilier to achieve a further refinement of our problem. Galileo selects and adapts an argument from his new kinematics—that is, he draws an argument from a set of arguments available for explanatory purposes, a set which I shall call the *explanatory store*. We may think of the sciences not as providing us with many unrelated individual arguments which can be used in individual acts of explanation, but as offering a reserve of explanatory arguments, which we may tap as need arises. Approaching the issue in this way, we shall be led to present our problem as that of specifying the conditions which must be met by the explanatory store.

The set of arguments which science supplies for adaptation acts of explanation will change with our changing beliefs. Therefore the appropriate *analysandum* is the notion of the store of arguments relative to a set of accepted sentences. Suppose that, at the point in the history of inquiry which interests us, the set of accepted sentences is K. (I shall assume, for simplicity's sake, that K is consistent. Should our beliefs be inconsistent then it is more appropriate to regard K as some tidied version of our beliefs.) The general problem I have set is that of specifying $E(K)$, the *explanatory store over K,* which is the set of arguments acceptable as the basis for acts of explanation by those whose beliefs are exactly the members of K. (For the purposes of this paper I shall assume that, for each K there is exactly one $E(K)$.)

The unofficial view answers the problem: for each K, $E(K)$ is the set of arguments which best unifies K. My task is to articulate the answer. I begin by looking at two historical episodes in which the desire for unification played a crucial role. In both cases, we find three important features: (i) prior to the articulation of a theory with high predictive power, certain proposals for theory construction are favored on grounds of their explanatory promise; (ii) the explanatory power of embryonic theories is explicitly tied to the notion of unification; (iii) particular features of the theories are taken to support their claims to unification. Recognition of (i) and (ii) will illustrate points that have already been made, while (iii) will point towards an analysis of the concept of unification.

3. A Newtonian Program

Newton's achievements in dynamics, astronomy and optics inspired some of his successors to undertake an ambitious program which I shall call "dynamic corpuscularianism."[4] *Principia* had shown how to obtain the motions of bodies from a knowledge of the forces acting on them, and had also demonstrated the possibility of dealing with gravitational systems in a unified way. The next step would be to isolate a few basic force laws, akin to the law of universal gravitation, so that, applying the basic laws to specifications of the dispositions of the ultimate parts of bodies, all of the phenomena of nature could be derived. Chemical reactions, for example, might be understood in terms of the rearrangement of ultimate parts under the action of cohesive and repulsive forces. The phenomena of reflection, refraction and diffraction of light might be viewed as resulting from a special force of attraction between light corpuscles and ordinary matter. These speculations encouraged eighteenth-century Newtonians to construct very general hypotheses about inter-atomic forces—even in the absence of any confirming evidence for the existence of such forces.

In the preface to *Principia,* Newton had already indicated that he took dynamic corpuscularianism to be a program deserving the attention of the scientific community:

> I wish we could derive the rest of the phenomena of Nature by the same kind of reasoning from mechanical principles, for I am induced by many reasons to suspect that they may all depend upon certain forces by which the particles of bodies, by some causes hitherto unknown, are either mutually impelled towards one another, and cohere in regular figures, or are repelled and recede from one another (Newton 1962, p. xviii. See also Newton 1952, pp. 401–2).

This, and other influential passages, inspired Newton's successors to try to complete the unification of science by finding further force laws analogous to the law of universal gravitation. Dynamic corpuscularianism remained popular so long as there was promise of significant unification. Its appeal began to fade only when repeated attempts to specify force laws were found to invoke so many different (apparently incompatible) attractive and repulsive forces that the goal of unification appeared unlikely. Yet that goal could still motivate renewed efforts to implement the program. In the second half of the eighteenth-century Boscovich revived dynamic corpuscularian hopes by claiming that the whole of natural philosophy can be reduced to "one law of forces existing in nature."[5]

The passage I have quoted from Newton suggests the nature of the unification that was being sought. *Principia* had exhibited how one style of argument, one "kind of reasoning from mechanical principles," could be used in the derivation of descriptions of many, diverse, phenomena. The unifying power of Newton's work consisted in its demonstration that one *pattern* of argument could be used again and again in the derivation of a wide range of accepted sentences. (I shall give a representation of the Newtonian pattern in Section 5.) In searching for force laws analogous to the law of universal gravitation, Newton's successors were trying to generalize the pattern of argument presented in *Principia,* so that one "kind of reasoning" would suffice to derive all phenomena of motion. If, furthermore, the facts studied by chemistry, optics, physiology and so forth, could be related to facts about particle motion, then one general pattern of argument would be used in the derivation of all phenomena. I suggest that this is the ideal of unification at which Newton's immediate successors aimed, which came to seem less likely to be attained as the eighteenth century wore on, and which Boscovich's work endeavored, with some success, to reinstate.

4. The Reception of Darwin's Evolutionary Theory

The picture of unification which emerges from the last section may be summarized quite simply: a theory unifies our beliefs when it provides one (or more generally, a few) pattern(s) of argument which can be used in the derivation of a large number of sentences which we accept. I shall try to develop this idea more precisely in later sections. But first I want to show how a different example suggests the same view of unification.

In several places, Darwin claims that his conclusion that species evolve through natural selection should be accepted because of its explanatory power, that ". . . the doctrine must sink or swim according as it groups and explains phenomena" (F. Darwin 1887; vol. 2, p. 155, quoted in Hull 1974, p. 292). Yet, as he often laments, he is unable to provide any complete derivation of any biological phenomenon—our ignorance of the appropriate facts and regularities is "profound." How, then, can he contend that the primary virtue of the new theory is its explanatory power?

The answer lies in the fact that Darwin's evolutionary theory promises to unify a host of biological phenomena (C. Darwin 1964, pp. 243–44). The eventual unification would consist in derivations of descriptions of these phenomena which would instantiate a common pattern. When Darwin expounds his doctrine what he offers us is the pattern. Instead of detailed explanations of the presence of some particular trait in some particular species, Darwin presents two "imaginary examples" (C. Darwin 1964, pp. 90–96) and a diagram, which shows, in a general way, the evolution of species *represented by schematic letters* (1964, pp. 116–26). In doing so, he exhibits a pattern of argument, which, he maintains, can be instantiated, *in principle*, by a complete and rigorous derivation of descriptions of the characteristics of any current species. The derivation would employ the principle of natural selection—as well as premises describing ancestral forms and the nature of their environment and the (unknown) laws of variation and inheritance. In place of detailed evolutionary stories, Darwin offers *explanation-sketches*. By showing how a particular characteristic would be advantageous to a particular species, he indicates an explanation of the emergence of that characteristic in the species, suggesting the outline of an argument instantiating the general pattern.

From this perspective, much of Darwin's argumentation in the *Origin* (and in other works) becomes readily comprehensible. Darwin attempts to show how his pattern can be applied to a host of biological phenomena. He claims that, by using arguments which instantiate the pattern, we can account for analogous variations in kindred species, for the greater variability of specific (as opposed to generic) characteristics, for the facts about geographical distribution, and so forth. But he is also required to resist challenges that the pattern cannot be applied in some cases, that premises for arguments instantiating the pattern will not be forthcoming. So, for example, Darwin must show how evolutionary stories, fashioned after his pattern, can be told to account for the emergence of complex organs. In both aspects of his argument, whether he is responding to those who would limit the application of his pattern or whether he is campaigning for its use within a realm of biological phenomena, Darwin has the same goal. He aims to show that his theory should be accepted because it unifies and explains.

5. Argument Patterns

Our two historical examples[6] have led us to the conclusion that the notion of an argument pattern is central to that of explanatory unification. Quite different considerations could easily have pointed us in the same direction. If someone were to distinguish between the explanatory worth of two arguments instantiating a common pattern, then we would regard that person as an explanatory deviant. To grasp the concept of explanation is to see that if one accepts an argument as explanatory, one is thereby committed to accepting as explanatory other arguments which instantiate the same pattern.

To say that members of a set of arguments instantiate a common pattern is to recognize that the arguments in the set are similar in some interesting way. With different interests, people may fasten on different similarities, and may arrive at different notions of argument pattern. Our enterprise is to characterize the concept of argument pattern which plays a role in the explanatory activity of scientists.

Formal logic, ancient and modern, is concerned in one obvious sense with patterns of argument. The logician proceeds by isolating a small set of expressions (the logical vocabulary), considers the schemata formed from sentences by replacing with dummy letters all expressions which do not belong to this set, and tries to specify which sequences of these schemata are valid patterns of argument. The pattern of argument which is taught to students of Newtonian dynamics is not a pattern of the kind which interests logicians. It has instantiations with different logical structures. (A rigorous derivation of the equations of motion of different dynamical systems would have a logical structure depending on the number of bodies involved and the mathematical details of the integration.) Moreover, an argument can only instantiate the Newtonian pattern if particular *non*-logical terms, 'force,' 'mass' and 'acceleration,' occur in it in particular ways. However, the logician's approach can help us to isolate the notion of argument pattern which we require.

Let us say that a *schematic sentence* is an expression obtained by replacing some, but not necessarily all, the non-logical expressions occurring in a sentence with dummy letters. A set of *filling instructions* for a schematic sentence is a set of directions for replacing the dummy letters of the schematic sentence, such that, for each dummy letter, there is a direction which tells us how it should be replaced. A *schematic argument* is a sequence of schematic sentences. A *classification* for a schematic argument is a set of sentences which describe the inferential characteristics of the schematic argument: its function is to tell us which terms in the sequence are to be regarded as premises, which are to be inferred from which, what rules of inference are to be used, and so forth.

We can use these ideas to define the concept of a *general argument pattern*. A general argument pattern is a triple consisting of a schematic argument, a set of sets of filling instructions containing one set of filling instructions for each term of the schematic argument, and a classification for the schematic argument. A sequence of sentences instantiates the general argument pattern just in case it meets the following conditions:

(i) The sequence has the same number of terms as the schematic argument of the general argument pattern.

(ii) Each sentence in the sequence is obtained from the corresponding schematic sentence in accordance with the appropriate set of filling instructions.

(iii) It is possible to construct a chain of reasoning which assigns to each sentence the status accorded to the corresponding schematic sentence by the classification.

We can make these definitions more intuitive by considering the way in which they apply to the Newtonian example. Restricting ourselves to the basic pattern used in treating systems which contain one body (such as the pendulum and the projectile) we may represent the schematic argument as follows:

(1) The force on α is β.
(2) The acceleration of α is γ.
(3) Force = mass · acceleration.
(4) (Mass of α) · (γ) = β
(5) $\delta = \theta$

The filling instructions tell us that all occurrences of 'α' are to be replaced by an expression referring to the body under investigation; occurrences of 'β' are to be replaced by an algebraic expression referring to a function of the variable coordinates and of time; 'γ' is to be replaced by an expression which gives the acceleration of the body as a function of its coordinates and their time-derivatives (thus, in the case of a one-dimensional motion along the x-axis of a Cartesian coordinate system, 'γ' would be relaced by the expression 'd^2x/dt^2'); 'δ' is to be replaced by an expression referring to the variable coordinates of the body, and 'θ' is to be replaced by an explicit function of time, (thus the sentences which instantiate (5) reveal the dependence of the variable coordinates on time, and so provide specifications of the positions of the body in question throughout the motion). The classification of the argument tells us that (1)–(3) have the status of premises, that (4) is obtained from them by substituting identicals, and that (5) follows from (4) using algebraic manipulation and the techniques of the calculus.

Although the argument patterns which interest logicians are general argument patterns in the sense just defined, our example exhibits clearly the features which distinguish the kinds of patterns which scientists are trained to use. Whereas logicians are concerned to display all the schematic premises which are employed and to specify exactly which rules of inference are used, our example allows for the use of premises (mathematical assumptions) which do not occur as terms of the schematic argument, and it does not give a complete description of the way in which the route from (4) to (5) is to go. Moreover, our pattern does not replace all non-logical expressions by dummy letters. Because some non-logical expressions remain, the pattern imposes special demands on arguments which instantiate it. In a different way, restrictions are set by the instructions for replacing dummy letters. The patterns of logicians are very liberal in both these latter respects. The conditions for replacing dummy letters in Aristotelian syllogisms, or first-order schemata, require only that some letters be relaced with predicates, others with names.

Arguments may be similar either in terms of their logical structure or in terms of the non-logical vocabulary they employ at corresponding places. I think that the notion of similarity (and the corresponding notion of pattern) which is central to the explanatory activity of scientists results from a compromise in demanding these two kinds of similarity. I propose that scientists are interested in *stringent* patterns of argument, patterns which contain some non-logical expressions and which are fairly similar in terms of logical structure. The Newtonian pattern cited above furnishes a good example. Although arguments instantiating this pattern do not have exactly the same logical structure, the classification imposes conditions which ensure that there will be similarities in logical structure among such arguments. Moreover, the presence of the non-logical terms sets strict requirements on the instantiations and so ensures a different type of kinship among them. Thus, without trying to provide an exact analysis of the notion of stringency, we may suppose that the stringency of a pattern is determined by two different constraints: (1) the conditions on the substitution of expressions for dummy letters, jointly imposed by the presence of non-logical expressions in the pattern and by the filling instructions; and, (2) the conditions on the logical structure, imposed by the classification. If both conditions are relaxed completely then the notion of pattern degenerates so as to admit *any* argument. If both conditions are simultaneously made as strict as possible, then we obtain another degenerate case, a "pattern" which is its own unique instantiation. If condition (2) is tightened at the total expense of (1), we produce the logician's notion of pattern. The use of condition (1) requires that arguments instantiating a common pattern draw on a common non-logical vocabulary. We can glimpse here that ideal of unification through the use of a few theoretical concepts which the remarks of Hempel and Feigl suggest.

Ideally, we should develop a precise account of how these two kinds of similarity are weighted against one another. The best strategy for obtaining such an account is to see how claims about stringency occur in scientific discussions. But scientists do not make explicit assessments of the stringency of argument patterns. Instead they evaluate the ability of a theory to explain and to unify. The way to a refined account of stringency lies through the notions of explanation and unification.

6. Explanation as Unification

As I have posed it, the problem of explanation is to specify which set of arguments we ought to accept for explanatory purposes given that we hold certain sentences to be true. Obviously this formulation can encourage confusion: we must not think of a scientific community as *first* deciding what sentences it will accept and *then* adopting the appropriate set of arguments. The Newtonian and Darwinian examples should convince us that the promise of explanatory power enters into the modification of our beliefs. So, in proposing that $E(K)$ is a function of K, I do not mean to suggest that the acceptance of K must be temporally prior to the adoption of $E(K)$.

$E(K)$ is to be that set of arguments which best unifies K. There are, of course, usually many ways of deriving some sentences in K from others. Let us call a set of arguments which derives some members of K from other members of K a *systematization* of K. We may then think of $E(K)$ as the best systematization of K.

Let us begin by making explicit an idealization which I have just made tacitly. A set of arguments will be said to be *acceptable relative* to K just in case every argument in the set consists of a sequence of steps which accord with elementary valid rules of inference (deductive or inductive) and if every premise of every argument in the set belongs to K. When we are considering ways of systematizing K we restrict our attention to those sets of arguments which are acceptable relative to K. This is an idealization because we sometimes use as the basis of acts of explanation arguments furnished by theories whose principles we no longer believe. I shall not investigate this practice nor the considerations which justify us in engaging in it. The most obvious way to extend my idealized picture to accommodate it is to regard the explanatory store over K, as I characterize it here, as being supplemented with an extra class of arguments meeting the following conditions: (a) from the perspective of K, the premises of these arguments are approximately true; (b) these arguments can be viewed as approximating the structure of (parts of) arguments in $E(K)$; (c) the arguments are simpler than the corresponding arguments in $E(K)$. Plainly, to spell out these conditions precisely would lead into issues which are tangential to my main goal in this paper.

The moral of the Newtonian and Darwinian examples is that unification is achieved by using similar arguments in the derivation of many accepted sentences. When we confront the set of possible systematizations of K we should therefore attend to the *patterns* of argument which are employed in each systematization. Let us introduce the notion of a *generating set*: if Σ is a set of arguments then a generating set for Σ is a set of argument patterns Π such that each argument in Σ is an instantiation of some pattern in Π. A generating set for Σ will be said to be *complete with respect to K* if and only if every argument which is acceptable relative to K and which instantiates a pattern in Π belongs to Σ. In determining the explanatory store $E(K)$ we first narrow our choice to those sets of arguments which are acceptable relative to K, the systematizations of K. Then we consider, for each such set of arguments, the various generating sets of argument patterns which are complete with respect to K. (The importance of the requirement of completeness is to debar explanatory deviants who use patterns selectively.) Among these latter sets we select that set with the greatest unifying power (according to criteria shortly to be indicated) and we call the selected set the *basis* of the set of arguments in question. The explanatory store over K is that systematization whose basis does best by the criteria of unifying power.

This complicated picture can be made clearer, perhaps, with the help of a diagram.

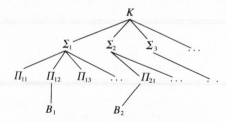

Systematizations, sets of arguments acceptable relative to K.

Complete generating sets. Π_{ij} is a generating set for Σ_i which is complete with respect to K.

Bases. B_i is the basis for Σ_i, and is selected as the best of the Π_{ij} on the basis of unifying power.

If B_k is the basis with the greatest unifying power then $E(K) = \Sigma_k$.

The task which confronts us is now formulated as that of specifying the factors which determine the unifying power of a set of argument patterns. Our Newtonian and Darwinian examples inspire an obvious suggestion: unifying power is achieved by generating a large number of accepted sentences as the conclusions of acceptable arguments which instantiate a few, stringent patterns. With this in mind, we define the *conclusion set* of a set of arguments Σ, $C(\Sigma)$, to be the set of sentences which occur as conclusions of some argument in Σ. So we might propose that the unifying power of a basis B_i with respect to K varies directly with the size of $C(\Sigma_i)$, varies directly with the stringency of the patterns which belong to B_i, and varies inversely with the number of members of B_i. This proposal is along the right lines, but it is, unfortunately, too simple.

The pattern of argument which derives a specification of the positions of bodies as explicit functions of time from a specification of the forces acting on those bodies is, indeed, central to Newtonian explanations. But not every argument used in Newtonian explanations instantiates this pattern. Some Newtonian derivations consist of an argument instantiating the pattern followed by further derivations from the conclusion. Thus, for example, when we explain why a pendulum has the period it does we may draw on an argument which *first* derives the equation of motion of the pendulum and *then* continues by deriving the period. Similarly, in explaining why projectiles projected with fixed velocity obtain maximum range when projected at 45° to the horizontal, we first show how the values of the horizontal and vertical coordinates can be found as functions of time and the angle of elevation, use our results to compute the horizontal distance traveled by the time the projectile returns to the horizontal, and then show how this distance is a maximum when the angle of elevation of projection is 45°. In both cases we take further steps beyond the computation of the explicit equations of motion—and the further steps in each case are different.

If we consider the entire range of arguments which Newtonian dynamics supplies for explanatory purposes, we find that these arguments instantiate a number of different patterns. Yet these patterns are not entirely distinct, for all of them proceed by using the computation of explicit equations of motion as a prelude to further derivation. It is natural to suggest that the pattern of computing equations of motion is the *core* pattern provided by Newtonian theory, and that the theory also shows how conclusions generated by arguments instantiating the core pattern can be used to derive further conclusions. In some Newtonian explanations, the core pattern is supplemented by a *problem-reducing pattern,* a pattern of argument which shows how to obtain a further type of conclusion from explicit equations of motion.

This suggests that our conditions on unifying power should be modified, so that, instead of merely counting the number of different patterns in a basis, we pay attention to similarities among them. All the patterns in the basis may contain a common core pattern, that is, each of them may contain some pattern as a subpattern. The unifying power of a basis is obviously increased if some (or all) of the patterns it contains share a common core pattern.

As I mentioned at the beginning of this [chapter], the account of explanation as unification is complicated. The explanatory store is determined on the basis of criteria which pull in different directions, and I shall make no attempt here to specify precisely the ways in which these criteria are to be balanced against one another. Instead, I shall show that some traditional problems of scientific explanation can be solved without more detailed specification of the conditions on unifying power. For the account I have indicated has two important corollaries.

(A)   Let Σ, Σ' be sets of arguments which are acceptable relative to K and which meet the following conditions:

> (i) the basis of Σ' is as good as the basis of Σ in terms of the criteria of stringency of patterns, paucity of patterns, presence of core patterns and so forth.
> (ii) $C(\Sigma)$ is a proper subset of $C(\Sigma')$.

Then $\Sigma \neq E(K)$.

(B)   Let Σ, Σ' be sets of arguments which are acceptable relative to K and which meet the following conditions:

> (i) $C(\Sigma) = C(\Sigma')$
> (ii) the basis of Σ' is a proper subset of the basis of Σ.

Then $\Sigma \neq E(K)$.

(A) and (B) tell us that sets of arguments which do equally well in terms of some of our conditions are to be ranked according to their relative ability to satisfy the rest. I shall try to show that (A) and (B) have interesting consequences.

7. Asymmetry, Irrelevance and Accidental Generalization

Some familiar difficulties beset the covering law model. The *asymmetry problem* arises because some scientific laws have the logical form of equivalences. Such laws can be used "in either direction." Thus a law asserting that the satisfaction of a condition C_1 is equivalent to the satisfaction of a condition C_2 can be used in two different kinds of argument. From a premise asserting that an object meets C_1, we can use the law to infer that it meets C_2; conversely, from a premise asserting that an object meets C_2, we can use the law to infer that it meets C_1. The asymmetry problem is generated by noting that in many such cases one of these derivations can be used in giving explanations while the other cannot.

Consider a hoary example. (For further examples, see Bromberger 1966.) We can explain why a simple pendulum has the period it does by deriving a specification of the period from a specification of the length and the law which relates length and period. But we cannot explain the length of the pendulum by deriving a specification of the length from a specification of the period and the same law. What accounts for our different assessment of these two arguments? Why does it seem that one is explanatory while the other "gets things backwards"? The covering law model fails to distinguish the two, and thus fails to provide answers.

The *irrelevance problem* is equally vexing. The problem arises because we can sometimes find a lawlike connection between an accidental and irrelevant occurrence and an event or state which would have come about independently of that occurrence. Imagine that Milo the magician waves his hands over a sample of table salt, thereby "hexing" it. It is true (and I shall suppose, lawlike) that all hexed samples of table salt dissolve when placed in water. Hence we can construct a derivation of the dissolving of Milo's hexed sample of salt by citing the circumstances of the hexing. Although this derivation fits the covering law model, it is, by our ordinary lights, non-explanatory. (This example is given by Wesley Salmon in his (1970); Salmon attributes it to Henry Kyburg. For more examples, see Achinstein 1971.)

The covering law model explicitly debars a further type of derivation which any account of explanation ought to exclude. Arguments whose premises contain no laws, but which make essential use of accidental generalizations are intuitively non-explanatory. Thus, if we derive the conclusion that Horace is bald from premises stating that Horace is a member of the Greenbury School Board and that all members of the Greenbury School Board are bald, we do not thereby explain why Horace is bald. (See Hempel 1965, p. 339.) We shall have to show that our account does not admit as explanatory derivations of this kind.

I want to show that the account of explanation I have sketched contains sufficient resources to solve these problems.[7] In each case we shall pursue a common strategy. Faced with an argument we want to exclude from the explanatory store we endeavor to show that any set of arguments containing the unwanted argument could not provide the best unification of our beliefs. Specifically, we shall try to show either that any such set of arguments will be more limited than some other set with an equally satisfactory basis, or that the basis of the set must fare worse according to the criterion of using the smallest number of most stringent patterns. That is, we shall appeal to the corollaries (A) and (B) given above. In actual practice, this strategy for exclusion is less complicated than one might fear, and, as we shall see, its applications to the examples just discussed brings out what is intuitively wrong with the derivations we reject.

Consider first the irrelevance problem. Suppose that we were to accept as explanatory the argument which derives a description of the dissolving of the salt from a description of Milo's act of hexing. What will be our policy for explaining the dissolving of samples of salt which have not been hexed? If we offer the usual chemical arguments in these latter cases then we shall commit ourselves to an inflated basis for the set of arguments we accept as explanatory. For, unlike the person who explains *all* cases of dissolving of samples of salt by using the standard chemical pattern of argument, we shall be committed to the use of two different patterns of argument in covering such cases. Nor is the use of the extra pattern of argument offset by its applicability in explaining other phenomena. Our policy employs one extra pattern of argument without extending the range of things we can derive from our favored set of arguments. Conversely, if we eschew the standard chemical pattern of argument (just using the pattern which appeals to the hexing) we shall find ourselves unable to apply our favored pattern to cases in which the sample of salt dissolved has not been hexed. Moreover, the pattern we use will not fall under the more general patterns we employ to explain chemical phenomena such as solution, precipitation and so forth. Hence the unifying power of the basis for our preferred set of arguments will be less than that of the basis for the set of arguments we normally accept as explanatory.[8]

If we explain the dissolving of the sample of salt which Milo has hexed by appealing to the hexing then we are faced with the problems of explaining the dissolving of unhexed samples of salt. We have two options: (a) to adopt two patterns of argument corresponding to the two kinds of case; (b) to adopt one pattern of argument whose instantiations apply just to the cases of hexed salt. If we choose (a) then we shall be in conflict with (B), whereas choice of (b) will be ruled out by

(A). The general moral is that appeals to hexing fasten on a local and accidental feature of the cases of solution. By contrast our standard arguments instantiate a pattern which can be generally applied.[9]

A similar strategy succeeds with the asymmetry problem. We have general ways of explaining why bodies have the dimensions they do. Our practice is to describe the circumstances leading to the formation of the object in question and then to show how it has since been modified. Let us call explanations of this kind "origin and development derivations." (In some cases, the details of the original formation of the object are more important; with other objects, features of its subsequent modification are crucial.) Suppose now that we admit as explanatory a derivation of the length of a simple pendulum from a specification of the period. Then we shall either have to explain the lengths of *non*-swinging bodies by employing quite a different style of explanation (an origin and development derivation) or we shall have to forego explaining the lengths of such bodies. The situation is exactly parallel to that of the irrelevance problem. Admitting the argument which is intuitively non-explanatory saddles us with a set of arguments which is less good at unifying our beliefs than the set we normally choose for explanatory purposes.

Our approach also solves a more refined version of the pendulum problem (given by Paul Teller in his (1974)). Many bodies which are not currently executing pendulum motion *could* be making small oscillations, and, were they to do so, the period of their motion would be functionally related to their dimensions. For such bodies we can specify the *dispositional period* as the period which the body would have if it were to execute small oscillations. Someone may now suggest that we can construct derivations of the dimensions of bodies from specifications of their dispositional periods, thereby generating an argument pattern which can be applied as generally as that instantiated in origin and development explanations. This suggestion is mistaken. There are some objects—such as the Earth and the Crab Nebula—which *could not* be pendulums, and for which the notion of a dispositional period makes no sense. Hence the argument pattern proposed cannot entirely supplant our origin and development derivations, and, in consequence, acceptance of it would fail to achieve the best unification of our beliefs.

The problem posed by accidental generalizations can be handled in parallel fashion. We have a general pattern of argument, using principles of physiology, which we apply to explain cases of baldness. This pattern is generally applicable, whereas that which derives ascriptions of baldness using the principle that all members of the Greenbury School Board are bald is not. Hence, as in the other cases, sets which contain the unwanted derivation be ruled out by one of the conditions (A), (B).

Of course, this does not show that an account of explanation along the lines I have suggested would sanction only derivations which satisfy the conditions imposed by the covering law model. For I have not argued that an explanatory derivation need contain *any* sentence of universal form. What *does* seem to follow from the account of explanation as unification is that explanatory arguments must not use accidental generalization, and, in this respect, the new account appears to underscore and generalize an important insight of the covering law model. Moreover, our success with the problems of asymmetry and irrelevance indicates that, even in the absence of a detailed account of the notion of stringency and of the way in which generality of the consequence set is weighed against paucity and stringency of the patterns in the basis, the view of explanation as unification has the resources to solve some traditional difficulties for theories of explanation.

8. Spurious Unification

Unfortunately there is a fly in the ointment. One of the most aggravating problems for the covering law model has been its failure to exclude certain types of self-explanation. (For a classic source of difficulties see Eberle, Kaplan and Montague 1961.) As it stands, the account of explanation as unification seems to be even more vulnerable on this score. The problem derives from a phenomenon which I shall call *spurious unification*.

Consider, first, a difficulty which Hempel and Oppenheim noted in a seminal article (Hempel 1965, Chapter 10). Suppose that we conjoin two laws. Then we can derive one of the laws from the conjunction, and the derivation conforms to the covering law model (unless, of course, the model is restricted to cover only the explanation of singular sentences; Hempel and Oppenheim do, in fact, make this restriction). To quote Hempel and Oppenheim:

> The core of the difficulty can be indicated briefly by reference to an example:
> Kepler's laws, K, may be conjoined with Boyle's law, B, to a stronger law $K \cdot B$;
> but derivation of K from the latter would not be considered as an explanation of
> the regularities stated in Kepler's laws; rather it would be viewed as representing,
> in effect, a pointless "explanation" of Kepler's laws by themselves. (Hempel
> 1965, p. 273, fn. 33.)

This problem is magnified for our account. For, why may we not unify our beliefs *completely* by deriving all of them using arguments which instantiate the one pattern?

$$\frac{\alpha \text{ and } B}{\alpha} \qquad [\text{`}\alpha\text{' is to be replaced by any sentence we accept.}]$$

Or, to make matters even more simple, why should we not unify our beliefs by using the most trivial pattern of self-derivation?

$$\alpha \qquad [\text{`}\alpha\text{' is to be replaced by any sentence we accept.}]$$

There is an obvious reply. The patterns just cited may succeed admirably in satisfying our criteria of using a few patterns of argument to generate many beliefs, but they fail dismally when judged by the criterion of stringency. Recall that the stringency of a pattern is assessed by adopting a compromise between two constraints: stringent patterns are not only to have instantiations with similar logical structures; their instantiations are also to contain similar non-logical vocabulary at similar places. Now both of the above argument patterns are very lax in allowing any vocabulary whatever to appear in the place of 'α'. Hence we can argue that, according to our intuitive concept of stringency, they should be excluded as non-stringent.

Although this reply is promising, it does not entirely quash the objection. A defender of the unwanted argument patterns may *artificially* introduce restrictions on the pattern to make it more stringent. So, for example, if we suppose that one of *our* favorite patterns (such as the Newtonian pattern displayed above) is applied to generate conclusions meeting a particular condition C, the defender of the patterns just cited may propose that 'α' is to be replaced, not by any sentence, but by a sentence which meets C. He may then legitimately point out that his newly contrived pattern is as stringent as our favored pattern. Inspired by this partial success, he may adopt a general strategy. Wherever we use an argument pattern to generate a particular type of conclusion, he may use some argument pattern which involves self-derivation, placing an appropriate restriction on the sentences to be substituted for the dummy letters. In this way, he will mimic whatever unification we achieve. His "unification" is obviously spurious. How do we debar it?

The answer comes from recognizing the way in which the stringency of the unwanted patterns was produced. Any condition on the substitution of sentences for dummy letters would have done equally well, provided only that it imposed constraints comparable to those imposed by acceptable patterns. Thus the stringency of the restricted pattern seems accidental to it. This accidental quality is exposed when we notice that we can vary the filling instructions, while retaining the same syntactic structure, to obtain a host of other argument patterns with equally many instantiations. By contrast, the constraints imposed on the substitution of non-logical vocabulary in the Newtonian pattern (for example) cannot be amended without destroying the stringency of the pattern or without depriving it of its ability to furnish us with many instantiations. Thus the constraints imposed in the Newtonian pattern are essential to its functioning; those imposed in the unwanted pattern are not.

Let us formulate this idea as an explicit requirement. If the filling instructions associated with a pattern P could be replaced by different filling instructions, allowing for the substitution of a class of expressions of the same syntactic category, to yield a pattern P' and if P' would allow the derivation *any* sentence, then the unification achieved by P is spurious. Consider, in this light, any of the patterns which we have been trying to debar. In each case, we can vary the filling instructions to produce an even more "successful" pattern. So, for example, given the pattern:

α ['α' is to be replaced by a sentence meeting condition C]

we can generalize the filling instructions to obtain

α ['α' is to be replaced by any sentence].

Thus, under our new requirement, the unification achieved by the original pattern is spurious.

In a moment I shall try to show how this requirement can be motivated, both by appealing to the intuition which underlies the view of explanation as unification and by recognizing the role that something like my requirement has played in the history of science. Before I do so, I want to examine a slightly different kind of example which initially appears to threaten my account. Imagine that a group of religious fanatics decides to argue for the explanatory power of some theological doctrines by claiming that these doctrines unify their beliefs about the world. They suggest that their beliefs can be systematized by using the following pattern:

God wants it to be the case that α. What God wants to be the case is the case.	['α' is to be replaced by any accepted sentence describing the physical world]

$$\frac{\text{God wants it to be the case that } \alpha.\ \text{What God wants to be the case is the case.}}{\alpha}$$

The new requirement will also identify as spurious the pattern just presented, and will thus block the claim that the theological doctrines that God exists and has the power to actualize his wishes have explanatory power. For it is easy to see that we can modify the filling instructions to obtain a pattern that will yield any sentence whatsoever.

Why should patterns whose filling instructions can be modified to accommodate any sentence be suspect? The answer is that, in such patterns, the non-logical vocabulary which remains is idling. The presence of that non-logical vocabulary imposes no constraints on the expressions we can substitute for the dummy symbols, so that, beyond the specification that a place be filled by expressions of a particular syntactic category, the structure we impose by means of filling instructions is quite incidental. Thus the patterns in question do not genuinely reflect the contents of our beliefs. The explanatory store should present the order of natural phenomena which is exposed by what we think we know. To do so, it must exhibit connections among our beliefs beyond those which could be found among any beliefs. Patterns of self-derivation and the type of pattern exemplified in the example of the theological community merely provide trivial, omnipresent connections, and, in consequence, the unification they offer is spurious.

My requirement obviously has some kinship with the requirement that the principles put forward in giving explanations be testable. As previous writers have insisted that genuine explanatory theories should not be able to cater to all possible evidence, I am demanding that genuinely unifying patterns should not be able to accommodate all conclusions. The requirement that I have proposed accords well with some of the issues which scientists have addressed in discussing the explanatory merits of particular theories. Thus several of Darwin's opponents complain that the explanatory benefits claimed for the embryonic theory of evolution are illusory, on the grounds that the style of reasoning suggested could be adapted to any conclusion. (For a particularly acute statement of the complaint, see the review by Fleeming Jenkin, printed in Hull 1974, especially p. 342.) Similarly, Lavoisier denied that the explanatory power of the phlogiston theory was genuine, accusing that theory of using a type of reasoning which could adapt itself to any conclusion (Lavoisier 1862, vol. II, p. 233). Hence I suggest that some problems of spurious unification can be solved in the way I have indicated, and that the solution conforms both to our intuitions about explanatory unification and to the considerations which are used in scientific debate.

However, I do not wish to claim that my requirement will debar all types of spurious unification. It may be possible to find other unwanted patterns which circumvent my requirement. A full characterization of the notion of a stringent argument pattern should provide a criterion for excluding the unwanted patterns. My claim in this section is that it will do so by counting as spurious the unification achieved by patterns which adapt themselves to any conclusion and by patterns which accidentally restrict such universally hospitable patterns. I have also tried to show how this claim can be developed to block the most obvious cases of spurious unification.

9. Conclusions

I have sketched an account of explanation as unification, attempting to show that such an account has the resources to provide insight into episodes in the history of science and to overcome some traditional problems for the covering law model. In conclusion, let me indicate very briefly how my view of explanation as unification suggests how scientific explanation yields understanding. By using a few patterns of argument in the derivation of many beliefs we minimize the number of *types* of premises we must take as underived. That is, we reduce, in so far as possible, the number of types of facts we must accept as brute. Hence we can endorse something close to Friedman's view of the merits of explanatory unification (Friedman 1974, pp. 18–19).

Quite evidently, I have only *sketched* an account of explanation. To provide precise analyses of the notions I have introduced, the basic approach to explanation offered here must be refined against concrete examples of scientific practice. What needs to be done is to look closely at the argument patterns favored by scientists and attempt to understand what characteristics they share. If I am right, the scientific search for explanation is governed by a maxim, once formulated succinctly by E. M. Forster. Only connect.

Notes

A distant ancestor of this paper was read to the Dartmouth College Philosophy Colloquium in the Spring of 1977. I would like to thank those who participated, especially Merrie Bergmann and Jim Moor, for their helpful suggestions. I am also grateful to two anonymous referees for *Philosophy of Science* whose extremely constructive criticisms have led to substantial improvements. Finally, I want to acknowledge the amount I have learned from the writing and the teaching of Peter Hempel. The present essay is a token payment on an enormous debt.

1. Strictly speaking, this is one of two views which emerge from Achinstein's discussion and which he regards as equally satisfactory. As Achinstein goes on to point out, either of these ontological theses can be developed to capture the central idea of the covering law model.

2. To pose the problem in this way we may still invite the charge that *arguments* should not be viewed as the bases for acts of explanation. Many of the criticisms leveled against the covering law model by Wesley Salmon in his seminal paper on statistical explanation (Salmon 1970) can be reformulated to support this charge. My discussion in section 7 will show how some of the difficulties raised by Salmon for the covering law model do not bedevil my account. However, I shall not respond directly to the points about statistical explanation and statistical inference advanced by Salmon and by Richard Jeffrey in his (1970). I believe that Peter Railton has shown how these specific difficulties concerning statistical explanation can

be accommodated by an approach which takes explanations to be (or be based on) arguments (see Railton 1978), and that the account offered in section 4 of his paper can be adapted to complement my own.

3. Of course, in restricting my attention to why-questions I am following the tradition of philosophical discussion of scientific explanation: as Bromberger notes in section IV of his (1966) not all explanations are directed at why-questions, but attempts to characterize explanatory responses to why-questions have a special interest for the philosophy of science because of the connection to a range of methodological issues. I believe that the account of explanation offered in the present paper could be extended to cover explanatory answers to some other kinds of questions (such as how-questions). But I do want to disavow the claim that unification is relevant to all types of explanation. If one believes that explanations are sometimes offered in response to what-questions (for example), so that it is correct to talk of someone explaining what a gene is, then one should allow that some types of explanation can be characterized independently of the notions of unification or of argument. I ignore these kinds of explanation in part because they lack the methodological significance of explanations directed at why-questions and in part because the problem of characterizing explanatory answers to what-questions seems so much less recalcitrant than that of characterizing explanatory answers to why-questions (for a similar assessment, see Belnap and Steel 1976, pp. 86–87). Thus I would regard a full account of explanation as a heterogeneous affair, because the conditions required of adequate answers to different types of questions are rather different, and I intend the present essay to make a proposal about how *part* of this account (the most interesting part) should be developed.

4. For illuminating accounts of Newton's influence on eighteenth-century research see Cohen (1956) and Schofield (1969). I have simplified the discussion by considering only *one* of the programs which eighteenth-century scientists derived from Newton's work. A more extended treatment would reveal the existence of several different approaches aimed at unifying science, and I believe that the theory of explanation proposed in this paper may help in the historical task of understanding the diverse aspirations of different Newtonians. (For the problems involved in this enterprise, see Heimann and McGuire 1971.)

5. See Boscovich (1966) Part III, especially p. 134. For an introduction to Boscovich's work, see the essays by L. L. Whyte and Z. Markovic in Whyte (1961). For the influence of Boscovich on British science, see the essays of Pearce Williams and Schofield in the same volume, and Schofield (1969).

6. The examples could easily be multiplied. I think it is possible to understand the structure and explanatory power of such theories as modern evolutionary theory, transmission genetics, plate tectonics, and sociobiology in the terms I develop here.

7. More exactly, I shall try to show that my account can solve some of the principal versions of these difficulties which have been used to discredit the covering law model. I believe that it can also overcome more refined versions of the problems than I consider here, but to demonstrate that would require a more lengthy exposition.

8. There is an objection this line of reasoning. Can't we view the arguments $\langle (x)((Sx$ and $Hx) \rightarrow Dx), Sa$ and $Ha, Da \rangle$, $\langle (x)((Sx$ and $\sim Hx) \rightarrow Dx), Sb$ and $\sim Hb, Db \rangle$ as instantiating a common pattern? I reply that, insofar as we can view these arguments as instantiating a common pattern, the standard pair of comparable (low-level) derivations—$\langle (x)(Sx \rightarrow Dx), Sa, Da \rangle$, $\langle (x)(Sx \rightarrow Dx), Sb, Db \rangle$—share a more stringent common pattern. Hence incorporating the deviant derivations in the explanatory store would give us an inferior basis. We can justify the claim that the pattern instantiated by the standard pair of derivations is more stringent than that shared by the deviant derivations, by noting that representation of the deviant pattern would compel us to broaden our conception of schematic sentence, and, even were we to do so, the deviant pattern would contain a "degree of freedom" which the standard pattern lacks. For a representation of the deviant "pattern" would take the form $\langle (x)((Sx$ and $\alpha Hx) \rightarrow Dx), Sa$ and $\alpha Ha, Da \rangle$, where 'α' is to be replaced uniformly either with the null

symbol or with '~'. Even if we waive my requirement that, in schematic sentences, we substitute for *non*-logical vocabulary, it is evident that this "pattern" is more accommodating than the standard pattern.

9. However, the strategy I have recommended will not avail with a different type of case. Suppose that a deviant wants to explain the dissolving of the salt by appealing to some property which holds universally. That is, the "explanatory" arguments are to begin from some premise such as "$(x)((x$ is a sample of salt and x does not violate conservation of energy) $\rightarrow x$ dissolves in water)" or "$(x)((x$ is a sample of salt and $x = x) \rightarrow x$ dissolves in water)." I would handle these cases somewhat differently. If the deviant's explanatory store were to be as unified as our own, then it would contain arguments corresponding to ours in which a redundant conjunct systematically occurred, and I think it would be plausible to invoke a criterion of simplicity to advocate dropping that conjunct.

References

Achinstein, P., *Law and Explanation*. Oxford: Oxford University Press 1971.

Achinstein, P., "What is an Explanation?," *American Philosophical Quarterly* 14(1977), pp. 1–15.

Belnap, N. and Steel, T. B., *The Logic of Questions and Answers*. New Haven: Yale University Press, 1976.

Boscovich, R. J., *A Theory of Natural Philosophy* (trans. J. M. Child). Cambridge: M.I.T. Press, 1966.

Bromberger, S., "An Approach to Explanation," in R. J. Butler (ed.), *Analytical Philosophy* (First Series). Oxford: Blackwell, 1962.

Bromberger, S., "Why-Questions," in R. Colodny (ed.), *Mind and Cosmos*. Pittsburgh: University of Pittsburgh Press, 1966.

Cohen, I. B., *Franklin and Newton*. Philadelphia: American Philosophical Society, 1956.

Darwin, C., *On the Origin of Species*, Facsimile of the First Edition, ed. E. Mayr. Cambridge: Harvard University Press, 1964.

Darwin, F., *The Life and Letters of Charles Darwin*. London: John Murray, 1987.

Eberle, R., Kaplan, D., and Montague, R., "Hempel and Oppenheim on Explanation," *Philosophy of Science* 28(1961), pp. 418–28.

Feigl, H., "The 'Orthodox' View of Theories: Remarks in Defense as Well as Critique," in M. Radner and S. Winokur (eds.), *Minnesota Studies in the Philosophy of Science*, vol. IV. Minneapolis: University of Minnesota Press, 1970.

Friedman, M., "Explanation and Scientific Understanding," *Journal of Philosophy*, vol. LXXI(1974), pp. 5–19.

Heimann, P., and McGuire, J. E., "Newtonian Forces and Lockean Powers," *Historical Studies in the Physical Sciences* 3(1971), pp. 233–306.

Hempel, C. G., *Aspects of Scientific Explanation*. New York: The Free Press, 1965.

———, "Deductive-Nonlogical vs. Statistical Explanation," in H. Feigl and G. Maxwell (eds.) *Minnesota Studies in the Philosophy of Science*, vol. III. Minneapolis: University of Minnesota Press, 1962.

———, *Philosophy of Natural Science*. Englewood Cliffs, N.J.: Prentice-Hall, 1966.

Hull, D. (ed.), *Darwin and his Critics*. Cambridge: Harvard University Press, 1974.

Jeffrey, R., "Statistical Explanation vs. Statistical Inference," in N. Rescher (ed.), *Essays in Honor of Carl G. Hempel*. Dordrecht: D. Reidel, 1970.

Kitcher, P. S., "Explanation, Conjunction and Unification," *Journal of Philosophy*, vol. LXXIII(1976), pp. 207–12.

Lavoisier, A., *Oeuvres*. Paris, 1862.

Newton, I., *The Mathematical Principles of Natural Philosophy* (trans. A. Motte and F. Cajori). Berkeley: University of California Press, 1962.

Newton, I., *Opticks*. New York: Dover, 1952.

Railton, P., "A Deductive-Nomological Model of Probabilistic Explanation," *Philosophy of Science* 45(1978), pp. 206–26.

Salmon, W., "Statistical Explanation," in R. Colodny (ed.), *The Nature and Function of Scientific Theories*. Pittsburgh: University of Pittsburgh Press, 1970.

Schofield, R. E., *Mechanism and Materialism*. Princeton: Princeton University Press, 1969.

Teller, P., "On Why-Questions," *Noûs,* vol. VIII(1974), pp. 371–80.

van Fraassen, B., "The Pragmatics of Explanation," *American Philosophical Quarterly* 14(1977), pp. 143–50.

Whyte, L. L. (ed.), *Roger Joseph Boscovich*. London: Allen and Unwin, 1961.

9

Explanation and Scientific Understanding

Michael Friedman

Why does water turn to steam when heated? Why do the planets obey Kepler's laws? Why is light refracted by a prism? These are typical of the questions science tries to answer. Consider, for example, the answer to the first question: Water is made of tiny molecules in a state of constant motion. Between these molecules are intermolecular forces, which, at normal temperatures, are sufficient to hold them together. If the water is heated, however, the energy, and consequently the motion, of the molecules increases. If the water is heated sufficiently the molecules acquire enough energy to overcome the intermolecular forces—they fly apart and escape into the atmosphere. Thus, the water gives off steam. This account answers our question. Our little story seems to give us understanding of the process by which water turns to steam. The phenomenon is now more intelligbile or comprehensible. How does this work? What is it about our little story, and scientific explanations generally, that gives us understanding of the world—what is it for a phenomenon to be scientifically understandable?

Two aspects of our example are of special interest. First, what is explained is a general regularity or pattern of behavior—a law, if you like—that is, that water turns to steam when heated. Although most of the philosophical literature on explanation deals with the explanation of particular events, the type of explanation illustrated by the above account seems much more typical of the physical sciences. Explanations of particular events are comparatively rare—found only perhaps in geology and astronomy. Second, our little story explains one phenomenon, the changing of water into steam, by relating it to another phenomenon, the behavior of the molecules of which water is composed. This relation is commonly described as *reduction*: the explained phenomenon is said to be reduced to the explaining phenomenon; for example, the behavior of water is reduced to the behavior of molecules. Thus, the central problem for the theory of scientific explanation comes down to this: what is the relation between phenomena in virtue of which one phenomenon can constitute

Michael Friedman, "Explanation and Scientific Understanding" in *The Journal of Philosophy*, Vol. 71, No. 1 (January 17, 1974), pp. 5–19. Reprinted by permission of *The Journal of Philosophy* and the author.

an explanation of another, and what is it about this relation that gives understanding of the explained phenomenon?

When I ask that a theory of scientific explanation tell us what it is about the explanation relation that produces understanding, I do not suppose that "scientific understanding" is a clear notion. Nor do I suppose that it is possible to say what scientific understanding is in advance of giving a theory of explanation. It is not reasonable to require that a theory of explanation proceed by first defining "scientific understanding" and then showing how its reconstruction of the explanation relation produces scientific understanding. We can find out what scientific understanding consists in only by finding out what scientific explanation is and vice versa. On the other hand, although we have no clear independent notion of scientific understanding, I think we do have some general ideas about what features such a notion should have, and we can use these general ideas to judge the adequacy of philosophical theories of explanation. At any rate, this is the method I will follow. I will argue that traditional accounts of scientific explanation result in notions of scientific understanding that have objectionable or counterintuitive features. From my discussion of traditional theories I will extract some general properties that a concept of scientific understanding ought to have. Finally, I will propose an account of scientific explanation that possesses these desirable properties.

It seems to me that the philosophical literature on explanation falls into two rough groups. Some philosophers, like Hempel and Nagel, have relatively precise proposals as to the nature of the explanation relation, but relatively little to say about the connection between their proposals and scientific understanding, that is, about what it is about the relation they propose that gives us understanding of the world. Other philosophers, like Toulmin, Scriven, and Dray, have a lot to say about understanding, but relatively vague ideas about just what relation it is that produces this understanding. To illustrate this situation I will discuss three attempts at explicating the explanation relation that have been prominent in the literature.

1. The best known philosophical account of explanation, the D-N model, is designed primarily as a theory about the explanation of particular events, but the view that the explanation relation is basically a deductive relation applies equally to our present concern. According to the D-N model, a description of one phenomenon can explain a description of a second phenomenon only if the first description entails the second. Of course, a deductive relation between two such descriptions is not sufficient for one to be an explanation of the other, as expounders of the D-N model readily admit.[1]

The entailment requirement puts a constraint on the explanation relation, but it does not by itself tell us what it is about the explanation relation that gives us understanding of the explained phenomenon, that makes the world more intelligible. In some of their writings, defenders of the D-N model give the impression that they consider such a task to lie outside the province of the philosopher of science, because concepts like 'understanding' and 'intelligibility' are psychological or pragmatic. For example, Hempel writes: "such expressions as 'realm of understanding' and 'comprehensible' do not belong to the vocabulary of logic, for they refer to psychological or pragmatic aspects of explanation."[2] He goes on to characterize the pragmatic aspects of explanation as those which vary from individual to individual. Explanation in its pragmatic aspects is "a relative notion, something can be significantly said to constitute an explanation in this sense only for this or that individual."[3] The philosopher of science, according to Hempel, aims at explicating the nonpragmatic aspects of explanation, the sense of 'explanation' on which *A* explains *B simpliciter* and not *for you or for me*.

I completely agree with Hempel's contention that the philosopher of science should be interested in an objective notion of explanation, a notion that doesn't vary capriciously from individual to individual. However, the considerations Hempel advances can serve as an argument against the attempt to connect explanation and understanding only by an equivocation on 'pragmatic'. In the sense in which such concepts as 'understanding', and 'comprehensible' are clearly pragmatic, 'pragmatic' means roughly the same as 'psychological', that is, having to do with the thoughts, beliefs, attitudes, etc. of persons. However, 'pragmatic' can also mean subjective as opposed to objective. In this sense, a pragmatic notion varies from individual to individual, and is therefore a relative notion. But a concept can be pragmatic in the first sense without being pragmatic in the second. Take the concept of rational belief, for example—presumably, if it is rational to believe a given sentence on given evidence it is so for anyone, and not merely for this or that individual. Similarly, although the notion of understanding, like knowledge and belief but unlike truth, just is a psychological notion, I don't see why it can't be a perfectly objective one. I don't see why there can't be an objective or rational sense of 'scientific understanding', a sense on which what is scientifically comprehensible is constant for a relatively large class of people. Therefore, I don't see how the philosopher of science can afford to ignore such concepts as 'understanding' and 'intelligibility' when giving a theory of the explanation relation.

Despite his reluctance, Hempel as a matter of fact does try to connect his model of explanation with the notion of understanding. He writes: "the [D-N] argument shows that, given the particular circumstances and the laws in question, the occurrence of the phenomenon *was to be expected*; and it is in this sense that the explanation enables us to *understand why* the phenomenon occurred."[4] Here, showing that a phenomenon was to be expected comes to this: if one had known "the particular circumstances and laws in question" before the explained phenomenon occurred, one would have had rational grounds for expecting the explained phenomenon to occur. The phenomenon would not have taken one by surprise.

This attempt at connecting explanation and understanding is clearly best suited to the special case of the explanation of particular events; for only particular events occur at definite times, and can thus actually be expected before their occurrence. Nevertheless, the account seems to fail even here, since understanding and rational expectation are quite distinct notions. To have grounds for rationally expecting some phenomenon is not the same as to understand it. I think that this contention is conclusively established by the well-known examples of prediction via so-called "indicator laws"—the barometer and the storm, Koplick spots and measles, etc. In these examples, one is able to predict some phenomenon on the basis of laws and initial conditions, but one has not thereby enhanced one's understanding of why the phenomenon occurred. To the best of my knowledge, Hempel himself accepts these counterexamples, and, because of them, would concede today that the D-N model provides at best necessary conditions for the explanation of particular events.

When we come to the explanation of general regularities or patterns of behavior the situation is even worse. Since general regularities do not occur at definite times, there is no question of literally expecting them. In this context, showing a phenomenon to have been expected can only mean having rational grounds for believing that the phenomenon does occur. And it is clear that having grounds for believing that a phenomenon occurs, though it may be part of understanding that phenomenon, is not sufficient for such understanding. Scientific explanations may involve the provision of grounds for believing that the explained phenomena occur, but it is not in virtue of the provision of such grounds that they give us understanding.

I conclude that the D-N model has the following advantages: It provides a clear, precise, and simple condition—entailment—that the explanation relation must satisfy, which, as a necessary condition, is not obviously mistaken. Also, it makes explanation relatively objective—what counts as an explanation does not depend on the arbitrary tastes of the scientist or the age. However, D-N theorists have not succeeded in saying what it is about the explanation relation that provides understanding of the world.

2. A second view, which has been surprisingly popular, holds that scientific explanations give us understanding of the world by relating (or reducing) unfamiliar phenomena to familiar ones. This view is inspired by such examples as the kinetic theory of gases, which, it is thought, gain their explanatory power by comparing unfamiliar phenomena, such as the Boyle-Charles law, to familiar phenomena, such as the movements of tiny billiard balls. P. W. Bridgman states this view very clearly: "I believe that examination will show that the essence of explanation consists in reducing a situation to elements with which we are so familiar that we accept them as a matter of course, so our curiosity rests."[5] Among contemporary writers, William Dray seems to favor this view:

> Why does the theory of geometrical optics explain the length of particular shadows? . . . it is surely because a ray diagram goes along with it, allowing us to think of light as travelling along ray lines, some of the lines passing over the wall and others coming to a dead halt on its surface. The shadow length is explained when . . . we think of light as "something travelling," that is, when we apply to it a very familiar and perhaps anthropomorphic way of thinking. . . .
> Thus the role of theory in such explanations is really *parasitic* upon the fact that it suggests, with the aid of postulated, unobservable, entities, a "hat-doffing" series of happenings which we are licensed to fill in . . . But if the travelling of observable entities along observable rails in a similar way would not explain a similar pattern of impact on encountering a wall, and if the jostling of a tightly packed crowd would not explain the straining and collapsing of the walls of a tent in which they were confined, then the corresponding scientific theories would not explain shadow lengths and the behavior of gases.[6]

The implication here is clearly that theories like the kinetic theory of gases are able to explain phenomena only to the extent that they relate them to more familiar processes and events.

This view, although it is initially attractive and does make an honest attempt to relate explanation and understanding, is rather obviously inadequate. As a matter of fact, many scientific explanations relate relatively familiar phenomena, such as the reflecting and refracting of light, to relatively unfamiliar phenomena, such as the behavior of electromagnetic waves. If the view under consideration were correct,

most of the explanations offered by contemporary physics, which postulate phenomena stranger and less familiar than any that they explain, could not possibly explain.
But, on reflection, it is not hard to see why this account fails so completely. For,
being familiar, just like being expected, is not at all the same thing as being understood. We are all familiar with the behavior of household appliances like radios,
televisions, and refrigerators; but how many of us understand why they behave the
way they do?

Michael Scriven, although he explicitly rejects the "familiarity" account of explanation, appears to hold a view which is similar to it in important respects. He
believes that in any given context each person possesses a "realm of understanding"—a set of phenomena which that person understands at a given time. The job
of explanation is to relate phenomena that are not in this privileged set to phenomena
that are in it:

> . . . the request for an explanation presupposes that *something* is understood,
> and a complete answer is one that relates the object of inquiry to the realm of
> understanding in some comprehensible and appropriate way. What this way is
> varies from subject matter to subject matter just as what makes something better
> than something else varies from the field of automobiles to solutions of chess
> problems, but the *logical function* of explanation, as of evaluation, is the same
> in each field.[7]

Thus, whereas on the "familiarity" view of the explanation relation the phenomenon
being explained must be related to a phenomenon that is familiar, on Scriven's account the phenomenon being explained must be related to a phenomenon that is
already understood. On both views the phenomenon doing the explaining must have
a special epistemological status. Not just any phenomenon will do. Both views conflict with the D-N account on this point. For, according to the D-N model, any
phenomenon (regardless of its familiarity or epistemological status) that bears the
appropriate deductive relation to the phenomenon being explained will do.

Scriven's view seems to me to be inadequate for the same reason as the "familiarity" view is. There just are many explanations in science which relate the
phenomena being explained to phenomena that are not themselves understood in the
relevant sense; that is, we do not understand why these latter phenomena occur. This
is true whenever a phenomenon is explained by reducing it to some "basic" or "fundamental" processes; for example, an explanation in terms of the behavior of the
fundamental particles of physics. In such cases the phenomenon doing the explaining
is not itself understood; it is simply a brute fact. But its ability to explain *other*
phenomena is not thereby impaired. Thus, I think that neither Scriven nor the "familiarity" theorists have given us good reason to suppose that it is a necessary feature
of the explanation relation that the phenomenon doing the explaining must itself have
some special epistemological status. It does not have to be a familiar or "hat doffing"
phenomenon, nor do we have to understand why *it* occurs. It merely has to explain
the phenomenon to which it is related.

3. A third approach to the explanation relation can be rather uncharitably labeled
the "intellectual fashion" view. Like the "familiarity" theorists, holders of this view
believe that the phenomenon doing the explaining must have a special epistemolog-

ical status, but, unlike the "familiarity" theorists, they think that this status varies from scientist to scientist and from historical period to historical period. At any given time certain phenomena are regarded as somehow self-explanatory or natural. Such phenomena need no explanation; they represent ideals of intelligibility. Explanation, within a particular historical tradition, consists in relating other phenomena to such ideals of intelligibility. Perhaps the clearest statement of this view is that of Stephen Toulmin, who calls such self-explanatory phenomena "ideals of natural order":

> . . . the scientist's prior expectations are governed by certain rational ideas or conceptions of the regular order of Nature. Things which happen according to these ideas he finds unmysterious; the cause or explanation of an event comes in question . . . through seemingly deviating from this regular way; its classi-fication among the different sorts of phenomena (for example, 'anomalous re-fraction') is decided by contrasting it with the regular, intelligible case; and, before the scientist can be satisfied, he must find some way of applying or ex-tending or modifying his prior ideas about Nature so as to bring the deviant event into the fold.[8]

Thus, the meaning of 'scientific understanding' varies with historical tradition, since what counts as an ideal of intelligibility does. Consequently, the very same theory may count as explanatory for one tradition but may fail to explain for another.

In all fairness, it should be pointed out that most supporters of this account do not believe that the choice of such ideals of intelligibility is completely capricious, depending only on the whims and prejudices of particular scientists. On the contrary, most believe that there can be good reasons, usually having to do with predictive power, for choosing one ideal over another. Indeed, one writer, N. R. Hanson, practically identifies predictive power with intelligibility. He argues that scientific theories typically go through three stages. When they are first proposed they are regarded as mere algorithms or "black boxes." As they begin to make more suc-cessful predictions than already existing theories, they become more respectable "grey boxes." Finally, through their ability to connect previously diverse areas of research, they become standards of intelligibility; they become what Hanson calls "glass boxes." The phenomena described by a theory in this third stage are taken as paradigms of naturalness and comprehensibility. According to Hanson, when a theory has suc-cessfully gone through these three stages "our very idea of what 'understanding' means will have grown and changed with the growth and change of the theory. So also will our idea of 'explanation'."[9]

This view clearly has a lot of historical support. There are many cases in the history of science where what seems explanatory to one scientist is a mere com-putational device for another; and there are cases where what is regarded as intel-ligible changes with tradition.[10] However, it seems to me that it would be desirable, if at all possible, to isolate a common, objective sense of explanation which remains constant throughout the history of science; a sense of 'scientific understanding' on which the theories of Newton, Maxwell, Einstein, and Bohr all produce scientific understanding. It would be desirable to find a concept of explanation according to which what counts as an explanation does not depend on what phenomena one finds particularly natural or self-explanatory. In fact, although there may be good reasons for picking one "ideal of natural order" over another, I cannot see any reason but prejudice for regarding some phenomena as somehow more natural, intelligible, or self-explanatory than others. All phenomena, from the commonest everyday event to the most abstract processes of modern physics, are equally in need of explanation—although it is impossible, of course, that they all be explained at once.

Therefore, although the "intellectual fashion" account may ultimately be the best that we can do, I don't see how it can give us what we are after: an objective and rational sense of 'understanding' according to which scientific explanations give us understanding of the world. We should try every means possible of devising an objective concept of explanation before giving in to something like the "intellectual fashion" account.

From the above discussion of three important contemporary theories of the explanation relation we can extract three desirable properties that a theory of explanation should have:

> 1. It should be sufficiently general—most, if not all, scientific theories that we all consider to be explanatory should come out as such according to our theory. This is where the "familiarity" theory fails, since, according to that view, theories whose basic phenomena are strange and unfamiliar—for example, all of contemporary physics—cannot be explanatory. Although it is unreasonable to demand that a philosophical account of explanation should show that every theory that has ever been thought to be explanatory really is explanatory, it must be at least square with most of the important, central cases.
>
> 2. It should be objective—what counts as an explanation should not depend on the idiosyncracies and changing tastes of scientists and historical periods. It should not depend on such nonrational factors as which phenomena one happens to find somehow more natural, intelligible, or self-explanatory than others. This is where the "intellectual fashion" account gives us less than we would like. If there is some objective and rational sense in which scientific theories explain, a philosophical theory of explanation should tell us what it is.
>
> 3. Our theory should somehow connect explanation and understanding—it should tell us what kind of understanding scientific explanations provide and how they provide it. This is where D-N theorists have been particularly negligent, although none of our three theories has given a satisfactory account of scientific understanding.

Thus, none of the three theories of explanation we have examined satisfies all our three conditions; none of them has succeeded in isolating a property of the explanation relation which is possessed by most of the clear, central cases of scientific explanation, which is common to the theories of scientists from various historical periods, and which has a demonstrable connection with understanding. Is there such a property?

Consider a typical scientific theory—for example, the kinetic theory of gases. This theory explains phenomena involving the behavior of gases, such as the fact that gases approximately obey the Boyle-Charles law, by reference to the behavior of the molecules of which gases are composed. For example, we can deduce that any collection of molecules of the sort that gases are, which obeys the laws of mechanics will also approximately obey the Boyle-Charles law. How does this make us understand the behavior of gases? I submit that if this were all the kinetic theory did we would have added nothing to our understanding. We would have simply replaced one brute fact with another. But this is not all the kinetic theory does—it also permits us to derive other phenomena involving the behavior of gases, such as the fact that they obey Graham's law of diffusion and (within certain limits) that they have the specific-heat capacities that they do have, from the same laws of mechanics. The kinetic theory effects a significant *unification* in what we have to accept. Where we once had three independent brute facts—that gases approximately obey the Boyle-Charles law, that they obey Graham's law, and that they have the specific-heat capacities they do have—we now have only one—that molecules obey the laws of mechanics. Furthermore, the kinetic theory also allows us to integrate the behavior of gases with other phenomena, such as the motions of the planets and of falling bodies near the earth. This is because the laws of mechanics also permit us to derive both the fact that planets obey Kepler's laws and the fact that falling bodies obey Galileo's laws. From the fact that *all* bodies obey the laws of mechanics it follows that the planets behave as they do, falling bodies behave as they do, and gases behave as they do. Once again, we have reduced a multiplicity of unexplained, independent phenomena to one. I claim that this is the crucial property of scientific theories we are looking for; this is the essence of scientific explanation— science increases our understanding of the world by reducing the total number of independent phenomena that we have to accept as ultimate or given. A world with fewer independent phenomena is, other things equal, more comprehensible than one with more.

Many philosophers have of course noticed the unifying effect of scientific theories to which I have drawn attention; for example, Hempel in one place writes: "a worthwhile scientific theory explains an empirical law by exhibiting it as one aspect of more comprehensive underlying regularities, which have a variety of other testable aspects as well, that is, which also imply various other empirical laws. Such a theory thus provides a systematically unified account of many different empirical laws." However, the only writer that I am aware of who has suggested that this unification or reduction in the number of independent phenomena is the essence of explanation in science is William Kneale:

> When we explain a given proposition we show that it follows logically from some other proposition or propositions. But this can scarcely be a complete account of the matter. . . . An explanation must in some sense simplify what we have to accept. Now the explanation of laws by showing that they follow from other laws is a simplification of what we have to accept because it reduces the number of untransparent necessitations we need to assume. . . . What we can achieve . . . is a reduction of the number of independent laws we need to assume for a complete description of nature.[11]

But does this idea really make sense? Can we give a clear meaning to "reduce the total number of independent phenomena"? Can we make our account a little less sketchy? First of all, I will suppose that we can represent what I have been calling *phenomena*—that is, general uniformities or patterns of behavior—by lawlike *sentences*; and that instead of speaking of the total number of independent phenomena we can speak of the total number of (logically) independent lawlike sentences. Secondly, since what is reduced is the total number of phenomena we have to accept, I will suppose that at any given time there is a set *K* of *accepted* lawlike sentences, a set of laws accepted by the scientific community. Furthermore, I will suppose that the set *K* is deductively closed in the following sense: if *S* is a lawlike sentence, and

$K \vdash S$, then S is a member of K; that is, K contains all lawlike consequences of members of K. Our problem now is to say when a given lawlike sentence permits a reduction of the number of independent sentences of K. For an example of what we want to characterize, let K contain the Boyle-Charles law, Graham's law, Galileo's law of free fall, and Kepler's laws, and let S be the conjunction of the laws of mechanics. Intuitively, we think S permits a reduction of the total number of independent sentences of K because we can replace a large number of independent laws by one (or at least by a smaller number); that is, we can replace the set containing the Boyle-Charles law, Graham's law, Galileo's law, and Kepler's laws by $\{S\}$. The trouble with this intuition is that it is not at all clear what counts as *one* law and what counts as *two*. For example, why haven't we reduced the number of independent laws if we replace the set containing the Boyle-Charles law, Graham's law, etc. by the unit set of their *conjunction*? It won't do to say that this conjunction is really not one law but four since it is logically equivalent to a set of four independent laws. For *any* sentence is equivalent to a set of n sentences for any finite n—for example, S is equivalent to $\{P, P \supset S\}$, where P is any consequence of S. I think the answer to this difficulty may be the following: although every sentence is equivalent to a set of n independent sentences, it is not the case that every sentence is equivalent to a set of n *independently acceptable* sentences—for example, the members of the set $\{P, P \supset S\}$ may not be acceptable independently of S; for our only grounds for accepting $P \supset S$, say, might be that it is a consequence of S.

I don't have anything very illuminating to say about what it is for one sentence to be acceptable independently of another. Presumably, it means something like: there are sufficient grounds for accepting one which are not also sufficient grounds for accepting the other. If this is correct, the notion of independent acceptability satisfies the following conditions:

(1) If $S \vdash Q$ then S is not acceptable independently of Q.

(2) If S is acceptable independently of P and $Q \vdash P$, then S is acceptable independently of Q.

(assuming that sufficient grounds for accepting S are also sufficient for accepting any consequence of S).

Given such a concept of independent acceptability, the notion of "reducing the number of independent sentences" can be made relatively precise in the following way. Let a *partition* of a sentence S be a set of sentences Γ such that Γ is logically equivalent to S and each S' in Γ is acceptable independently of S. Thus, if S is the conjunction of the Boyle-Charles law, Graham's law, Galileo's law, and Kepler's laws, the set Γ containing the conjuncts is a partition of S. I will say that a sentence S is *K-atomic* if it has no partition; that is, if there is no pair $\{S_1, S_2\}$ such that S_1 and S_2 are acceptable independently of S and S_1 & S_2 is logically equivalent to S. Thus, the above conjunction is not K-atomic. Let a *K-partition* of a set of sentences Δ be a set Γ of K-atomic sentences which is logically equivalent to Δ (I assume that such a K-partition exists for every set Δ). Let the *K-cardinality* of a set of sentences Δ, K-card (Δ), be inf $\{$card (Γ): Γ a K-partition of $\Delta\}$. Thus, if S is the above conjunction, K-card $(\{S\})$ is at least 4. Finally, I will say that S *reduces* the set Δ iff K-card $(\Delta \cup \{S\}) < K$-card (Δ). Thus, if S is the above conjunction and Γ is the set of its conjuncts, S does not reduce Γ.

How can we define *explanation* in terms of these ideas? If S is a candidate for explaining some S' in K, we want to know whether S permits a reduction in the number of independent sentences. I think that the relevant set we want S to reduce is the set of *independently acceptable* consequences of S ($con_K(S)$). For instance, Newton's laws are a good candidate for explaining Boyle's law, say, because Newton's laws reduce the set of their independently acceptable consequences—the set containing Boyle's law, Graham's law, etc. On the other hand, the *conjunction* of Boyle's law and Graham's law is not a good candidate, since it does not reduce the set of its independently acceptable consequences. This suggests the following definition of explanation between laws:

(D1) S_1 explains S_2 iff $S_2 \in con_K(S_1)$ and S_1 reduces $con_K(S_1)$

Actually this definition seems to me to be too strong; for if S_1 explains S_2 and S_3 is some independently acceptable law, then S_1 & S_3 will not explain S_2—since S_1 & S_3 will not reduce $con_K(S_1$ & $S_3)$. This seems undesirable—why should the conjunction of a completely irrelevant law to a good explanation destroy its explanatory power? So I will weaken (D1) to

(D1) S_1 explains S_2 iff there exists a partition Γ of S_1 and an $S_i \in \Gamma$ such that S_2 $\in con_K(S_i)$ and S_i reduces $con_K(S_i)$.

Thus, if S_1 explains S_2, then so does S_1 & S_3; for $\{S_1, S_3\}$ is a partition of S_1 & S_3, and S_1 reduces $con_K(S_1)$ by hypothesis.

Note that this definition is not vulnerable to the usual "conjunctive" trivialization of deductive theories of explanation alluded to in footnote 1 above; that is, the conjunction of two independent laws does not explain its conjuncts. Furthermore, my account shows why such a conjunction cannot be a good explanation. It does not increase our understanding since it does not reduce its independently acceptable consequences.

On the view of explanation that I am proposing, the kind of understanding provided by science is global rather than local. Scientific explanations do not confer intelligibility on individual phenomena by showing them to be somehow natural, necessary, familiar, or inevitable. However, our overall understanding of the world is increased; our total picture of nature is simplified via a reduction in the number of independent phenomena that we have to accept as ultimate. It seems to me that previous attempts at connecting explanation and understanding have failed through ignoring the global nature of scientific understanding. If one concentrates only on the local aspects of explanation—the phenomenon being explained, the phenomenon doing the explaining, and the relation (deductive or otherwise) between them—one ends up trying to find some special epistemological status—familiarity, naturalness, or being an "ideal of natural order"—for the phenomenon doing the explaining. For how else is understanding conferred on the phenomenon being explained? However, attention to the global aspects of explanation—the relation of the phenomena in question to the total set of accepted phenomena—allows one to dispense with any special epistemological status for the phenomenon doing the explaining. As long as a reduction in the total number of independent phenomena is achieved, the basic phenomena to which all others are reduced can be as strange, unfamiliar, and unnatural as you wish—even as strange as the basic facts of quantum mechanics.

This global view of scientific understanding also, it seems to me, provides the correct answer to the old argument that science is incapable of explaining anything because the basic phenomena to which others are reduced are themselves neither explained nor understood. According to this argument, science merely transfers our puzzlement from one phenomenon to another; it replaces one surprising phenomenon by another equally surprising phenomenon. The standard answer to this old argument is that phenomena are explained one at a time; that a phenomenon's being itself unexplained does not prevent it from explaining other phenomena in turn. I think this reply is not quite adequate. For the critic of science may legitimately ask how our total understanding of the world is increased by replacing one puzzling phenomenon with another. The answer, as I see it, is that scientific understanding is a global affair. We don't simply replace one phenomenon with another. We replace one phenomenon with a *more comprehensive* phenomenon, and thereby effect a reduction in the total number of accepted phenomena. We thus genuinely increase our understanding of the world.

Notes

An earlier version of this paper was read at the University of Massachusetts at Amherst. I am indebted to members of the philosophy department there, especially Fred Feldman, for helpful comments. I would also like to thank Hartry Field, Clark Glymour, and David Hills for valuable criticism and conversation.

1. Compare C. G. Hempel and P. Oppenheim. "Studies in the Logic of Explanation," in Hempel's *Aspects of Scientific Explanation* (New York: Free Press, 1965), p. 273, note 33; parenthetical page references to Hempel are to this article. The difficulty is that the conjunction of two laws entails each of its conjucts but does not necessarily explain them.

2. C. G. Hempel, "Aspects of Scientific Explanation" in *Aspects of Scientific Explanation and Other Essays in the Philosophy of Science*. New York: The Free Press, 1964, p. 413.

3. Ibid., p. 426.

4. Ibid., p. 327.

5. P. W. Bridgeman, *The Logic of Modern Physics* (New York: Macmillan, 1968), p. 37.

6. *Laws and Explanation in History* (New York: Oxford, 1961), pp. 79–80.

7. "Explanations, Predictions, and Laws," in H. Feigl and G. Maxwell, eds., *Minnesota Studies in the Philosophy of Science*, vol. III (Minneapolis: Univ. of Minnesota Press, 1970), p. 202.

8. Stephen Toulmin, *Foresight and Understanding* (New York: Harper & Row, 1963), pp. 45–46.

9. N. R. Hanson, *The Concept of the Positron* (New York: Cambridge, 1963), p. 38.

10. Examples can be found in Toulmin, Hanson, and T. Mischel, "Pragmatic Aspects of Explanation," *Philosophy of Science*, vol. XXXIII, no. 1 (March 1966): 40–60.

11. *Probability and Induction* (New York: Oxford, 1949), pp. 91–92.

10

The Illocutionary Theory
of Explanation

Peter Achinstein

Part I   Explaining

1. Conditions for an Act of Explaining

The verb "to explain" is, to borrow a classification from Zeno Vendler, an accomplishment term.[1] It has a continuous present, "is explaining," that indicates that an act is occurring that occupies some stretch of time. But unlike some other verbs which also have a continuous present, such as "to run" and "to push" (which Vendler calls activity terms), it has a past tense which indicates not simply a stop to the act but a conclusion or completion. If John was running, then no matter for how long he was running, he ran. But if the doctor was explaining Bill's stomach ache, then it is not necessarily true that he explained it, since his act may have been interrupted before completion.

Sylvain Bromberger suggests that although the accomplishment use of "explain" is the most fundamental there is also a non-accomplishment use, illustrated by saying that Newton explained the tides. Here

> one need not mean that some explaining episode took place in which Newton was the tutor. One may mean that Newton solved the problem, found the answer to the question.[2]

Using Vendler's terminology, Bromberger classifies this as an "achievement" use of "explains." Achievement terms (e.g., "winning a race"), unlike accomplishment terms, describe something that occurs at a single moment rather than over a stretch of time.

I believe that Bromberger is mistaken. If Newton has simply solved the problem or found the answer, although he may be in a position to explain the tides (we say that he *would* or *can* explain them as follows), he has not yet done so until he has said, or written, or at least communicated, something. One does not explain simply

From *The Nature of Explanation*, pp. 15–19, 74–94, and 98–102, by Peter Achinstein. Copyright © 1983 by Oxford University Press, Inc. Reprinted by permission.

by believing, or even by solving a problem or finding an answer, unless that belief, solution, or answer is expressed in some act of uttering or writing. (We sometimes explain rather simple things by non-verbal acts such as gesturing, but such cases will not be of concern to me here.) This does not mean that we must construe "Newton explained the tides" as describing a particular explaining episode. Following David-son,[3] this can be treated as an existentially general sentence, that is, as saying that there was at least one act which was an explaining of the tides by Newton.

Explaining is what Austin calls an illocutionary act.[4] Like warning and prom-ising, it is typically performed by uttering words in certain contexts with appropriate intentions. It is to be distinguished from what Austin calls perlocutionary acts, such as enlightening someone, or getting someone to understand, or removing someone's puzzlement, which are the effects one's act of explaining can have upon the thoughts and beliefs of others.

The illocutionary character of explaining can be exposed by formulating a set of conditions for performing such an act. To do so I shall consider sentences of the form "S explains q by uttering u," in which S denotes some person, q expresses an indirect question, and u is a sentence. (I will assume that any sentence of this form in which q is not an indirect question is transformable into one that is.)

The first condition expresses what I take to be a fundamental relationship be-tween explaining and understanding. It is that S explains q by uttering u only if

(1) S utters u with the intention that his utterance of u render q understandable.

This expresses the central point of S's act. It is the most important feature which distinguishes explaining from other illocutionary acts, even ones that can have in-direct questions as objects. If by uttering u I am asking you, or agreeing with you about, why the tides occur, by contrast to explaining it, I will not be doing so with the intention that my utterance render why the tides occur understandable. [. . .]

To explain q is not to utter just anything with the intention that the utterance render q understandable. Suppose I believe that the words "truth is beauty" are so causally efficacious with you that the mere uttering of them will cause you to un-derstand anything, including why the tides occur. By uttering these words I have not thereby explained why the tides occur, even if I have satisfied (1). The reason is that I do not believe that "truth is beauty" expresses a correct answer to the ques-tion "Why do the tides occur?" More generally, assuming that answers to questions are propositions [. . .] we may say that S explains q by uttering u only if

(2) S believes that u expresses a proposition that is a correct answer to Q. (Q is the direct form of the question whose indirect form is q.)

Often people will present hints, clues, or instructions which do not themselves an-swer the question but enable an answer to be found by others. To the question "Why do the tides occur?" I might respond: "Look it up in Chapter 10 of your physics text," or "Newton's *Principia* has the answer," or "Think of gravity." Some hints, no doubt, border on being answers to the question. But in those cases where they do not, it is not completely appropriate to speak of explaining. By uttering "Look it up in Chapter 10 of your physics text" I am not explaining why the tides occur, though I am uttering something which, I believe, will put you in a position to explain this.

These conditions are not yet sufficient. Suppose that S intends that his utterance of u render q understandable not by producing the knowledge that u expresses a correct answer to Q but by causing people to come to think of some non-equivalent sentence u' which, like u, S believes expresses a proposition that is a correct answer to Q. In such a case, although S utters something which he believes will cause others to be able to explain q, S does not himself explain q by uttering u. For example, to an audience that I believe already knows that the tides occur because of gravitational attraction, I say

> u: The tides occur because of gravitational attraction of the sort described by Newton.

Although I believe that u does express a correct answer to Q (Why do the tides occur?), suppose that I utter u with the following intention: that this utterance will render q understandable not by producing the knowledge of the proposition expressed by u that it is a correct answer to Q, but by causing my audience to look up the more detailed and precise answer actually supplied by Newton, which I don't present. This is like the situation in which I give the audience a hint that in this case is a correct answer, but is not the answer in virtue of which I intend q to be understandable to that audience.

To preclude such cases we can say that S explains q by uttering u only if

> (3) S utters u with the intention that his utterance of u render q understandable by producing the knowledge, of the proposition expressed by u, that it is a correct answer to Q.

In the case of the tides mentioned above, I do not intend that my utterance of u render q understandable by producing the knowledge *of the proposition expressed by u* that it is a correct answer to Q, but by producing such knowledge with respect to another proposition. So, according to condition (3), in such a case by uttering u I am not explaining why the tides occur.

Suppose, by contrast, I know that my audience is familiar with the answer supplied by Newton, but its members have no idea whether this answer is correct. Since the audience knows what sort of gravitational attraction Newton describes, I might explain why the tides occur, simply by uttering u. In this case I intend to render q understandable by producing the knowledge, of the proposition expressed by u, that it is a correct answer to Q. It is possible for me to have this intention with respect to u since I know that the audience is aware of the sort of gravitational attraction described by Newton.

Let us change the example once more. Suppose I believe that the audience does not know that the tides are due to gravitational attraction. I now proceed to utter u above with the intention that my utterance of u will render q understandable by the following combination of means (which I regard as jointly but not separately sufficient for rendering q understandable): (i) producing the knowledge, of the proposition expressed by u, that it is a correct answer to Q; and (ii) causing others to look up some different, more detailed, proposition (supplied by Newton) which is also a correct answer to Q. By uttering u am I explaining why the tides occur?

One might be inclined to say that I am *both* explaining q by uttering u *and* giving a clue about where to find another answer to Q. If this is correct, then (3) should be understood in a way that allows S to intend to render q understandable by a combination of means that includes producing the knowledge, of the proposition expressed by u, that it is a correct answer to Q. On the other hand, in the case just envisaged one might be tempted to say that I am doing something that falls between explaining and giving clues but is not exactly either. If this is correct, then (3) should be understood in a way that requires S to intend to render q understandable *solely* by producing the knowledge, of the proposition expressed by u, that it is a correct answer to Q. I am inclined to regard the latter interpretation of (3) as preferable, but I will not press the point. (This, of course, does not preclude S from explaining q by formulating a number of different propositions whose conjunction constitutes an answer to Q, or from engaging in several acts in which different, though not necessarily competing, answers to Q are provided.) [. . .] For the present I shall treat these three conditions as not only necessary but jointly sufficient. If so, then the same honor can be accorded to (3) by itself, since (3) entails both (1) and (2).

   Although "explain" may be used in describing an act governed by these conditions, it can also be employed in a more restricted way to cover only correct explainings. We can say that Galileo explained why the tides occur, even though he did so incorrectly, or that he failed to explain this, even though he tried. When one has correctly explained q by uttering u one has performed the illocutionary act of explaining q and in doing so one has provided a correct answer to Q. In what follows, however, when reference is made to acts of explaining I shall mean acts for which this is not a requirement.

Part II   What Is an Explanation?

Most theories of explanation focus not on explaining acts but on what I have been calling the products of such acts. They make claims about the ontological status of such products and about their evaluation. What sort of entity, if any, is an explanation? Views frequently expressed are that explanations are sentences, or propositions, or arguments. It is my aim here to examine these and other possibilities. I want to show why the simpler candidates will not do, and that if explanations are entities they are more complex than is generally believed. I also plan to ask whether explanation products can be understood independently of the concept of an explaining act. At the end I shall raise the question of how dependent on its ontology is a theory that provides conditions for evaluating explanations. Would criteria for evaluating explanations supplied by the D-N theory, for example, be precluded by the adoption of an ontology different from that proposed by the theory itself?

1. Sentence and Proposition Views

Let us suppose that some explaining act has occurred, for example, Dr. Smith explained why Bill got a stomach ache, by uttering the sentence

   (1) Bill ate spoiled meat.

Consider the explanation of why Bill got a stomach ache which is a product of Dr. Smith's explaining act. (In what follows I shall use product-expressions of the form "S's explanation of q" or "the explanation of q given by S" in the sense of "the explanation of q which is the product of S's act of explaining q.") To what sort of entity, if any, does the product-expression "Dr. Smith's explanation of why Bill got a stomach ache" refer?

A simple answer is that it refers to a sentence. Which one? Obviously, the sentence that the doctor uttered, viz. (1). In general, if S explains q by uttering a sentence, it might naturally be supposed that the product-expression "the explanation of q given by S" denotes the sentence S utters. Assuming that such a product-expression is uniquely referring, we can formulate the following condition:

Denotation condition (*sentence view*): "The explanation of q given by S" denotes u if and only if

    (i) u is a sentence;

    (ii) S explained q by uttering u;

    (iii) $(v)(S$ explained q by uttering $v \supset v = u)$. (That is, whatever S explained q by uttering is identical with u.)[5]

In accordance with this view, *explanations are sentences* (including conjunctions) *by the uttering of which explainers explain.* [. . .]

Two major problems beset the sentence theory, but since these are common to other views I shall deal with them later. However, a point will now be mentioned which suggests the preferability of a proposition view. Suppose that Dr. Smith explained Bill's stomach ache (why Bill got a stomach ache) by uttering

    (1) Bill ate spoiled meat,

while Dr. Robinson explained it by uttering

    (2) Bill ate meat that was spoiled.

If (1) is the only sentence by the uttering of which Dr. Smith explained Bill's stomach ache, and (2) is the only sentence by the uttering of which Dr. Robinson did so, then, on the present view, "the explanation of Bill's stomach ache given by Dr. Smith" denotes sentence (1), while "the explanation of Bill's stomach ache given by Dr. Robinson" denotes sentence (2). But since sentence (1) $\neq$ sentence (2),

the explanation of Bill's stomach ache given by Dr. Smith $\neq$

the explanation of Bill's stomach ache given by Dr. Robinson,

which seems unsatisfactory. Intuitively, both doctors have given the same explanation. Both have attributed the stomach ache to the eating of spoiled meat, despite the fact that the particular sentences used by each are not the same.

This multiplication of explanations can be avoided by identifying explanations with propositions, not sentences.[6] Since sentences (1) and (2) express the same proposition we can conclude that the explanations given by the doctors are the same. Let us then formulate the following

Denotation condition (*proposition view*): "The explanation of q given by S" denotes x if and only if

    (i) x is a proposition;

    (ii) $(\exists u)(S$ explained q by uttering $u)$;

    (iii) $(u)(S$ explained q by uttering $u \supset u$ expresses $x)$.

On this view *explanations are propositions expressed by sentences by the uttering of which explainers explain.*

Two considerations, however, make this (as well as the sentence view) untenable.

2. *The Illocutionary Force Problem*

To formulate the first, let me begin with a general observation about illocutionary acts and products, viz.

(1) The product of S's illocutionary act is an (illocutionary product) F only if S F-ed.

For example, the product of S's illocutionary act is a promise, or a warning, or a criticism, or an explanation, only if S promised, or warned, or criticized, or explained. This does not preclude there being several illocutionary products when S utters u, so long as, by uttering u, S is performing several illocutionary acts.

Let us suppose, now, that by uttering the sentence

(2) Bill ate spoiled meat

Jane criticized Bill for eating spoiled meat. Then

(3) The criticism of Bill given by Jane is that Bill ate spoiled meat.

By analogy with the proposition view of the product of explanation, the product-expression "the criticism of Bill given by Jane" will be taken to denote the proposition expressed by the sentence Jane uttered in giving her criticism, viz. the proposition expressed by (2).[7] On the proposition view, then, the product-expression in (3) denotes the same proposition as the product-expression "the explanation of Bill's stomach ache given by Dr. Smith" in our earlier example. Therefore, the explanation of Bill's stomach ache given by Dr. Smith = the criticism of Bill given by Jane. But the criticism of Bill given by Jane was a criticism. Therefore,

(4) The explanation of Bill's stomach ache given by Dr. Smith was a criticism.

Now in accordance with (1)—the general claim about illocutionary acts and products—the product of Dr. Smith's illocutionary act is a criticism, that is, (4) is true, only if Dr. Smith criticized. However, when Dr. Smith performed the illocutionary act of explaining Bill's stomach ache (we may suppose) he was not criticizing anyone. So (4) is false. What is true is only that what Dr. Smith uttered in explaining Bill's stomach ache is what Jane uttered in criticizing Bill. But (4) does not follow from that.

This will be called the illocutionary force problem. The proposition expressed by what is uttered in an act of explaining may be the same as the proposition expressed by what is uttered in other illocutionary acts, such as an act of criticizing. If the product-expression denotes such a proposition then we will have to conclude that the explanation given by S is a criticism, and hence that when S explained he criticized, even when this is not so. The proposition view (as well as the sentence view) is beset by the illocutionary force problem. It cannot distinguish explanations from the products of other illocutionary acts, where these products are not explanations.

To this someone might reply by rejecting assumption (1) regarding products of illocutionary acts. Suppose Jane in criticizing Bill utters "Bill ate spoiled meat." Dr. Smith hearing her utterance replies

(5) That is an explanation of Bill's stomach ache.

It would seem that (1) is violated since the product of Jane's illocutionary act is an explanation even though Jane was only criticizing and was not explaining anything. Has (1) really been violated in such a case? In response to (5) Jane can reply

(6) That is not an explanation of anything. (I didn't even know that Bill got a stomach ache.) It is simply a criticism of Bill—people shouldn't eat spoiled meat.

Suppose the tables are turned and Jane, hearing Dr. Smith utter "Bill ate spoiled meat," asserts

(7) That is criticism of Bill.

To this Dr. Smith replies

(8) That is not a criticism of anyone. It is simply an explanation of Bill's stomach ache.

I believe that both (5) and (6) (and (7) and (8)) are in the imagined contexts reasonable responses. The "dispute" here is more apparent than real since what is being referred to by "that" in (5) and (6) (and in (7) and (8)) is not the same. In (5) the doctor is referring to the sentence Jane uttered (or perhaps to the proposition it expresses). He is telling us, in effect, that this sentence (or proposition) can be used in providing an explanation of Bill's stomach ache. And this is perfectly true. In (6) Jane is referring to the product of her illocutionary act. She is telling us that this product is not an explanation but a criticism, since her act was a criticizing one, not an explaining one. And this is also true. If (5) is not referring to the product of an illocutionary act, whereas (6) is, then the general assumption (1) is not violated.

3. The Emphasis Problem

The second problem with the proposition view is that its denotation condition assumes that the u-position in sentences of the form "S explains q by uttering u" is referentially transparent when this position is filled by expressions for sentences. That is, it assumes that, from "S explains q by uttering u" and "$u = v$," we may infer "S explains q by uttering v." But this assumption can be seen to be unjustifiable by appeal to the notion of emphasis.[8] [. . .]

There are expressions for sentences which are such that adding or changing emphases in them will not alter the sentence referred to. For example, suppose I refer to

(1) The sentence "Bill ate spoiled meat on Tuesday,"

and you, who are hard of hearing, think I was referring to the sentence "Bill ate spoiled meat on Monday." I might reply that I am referring to

(2) The sentence "Bill ate spoiled meat *on Tuesday*."

By using emphasis in (2) I am not referring to a different sentence but to the same one as in (1). Only I use emphasis in (2) and not in (1) to correct your mistake about which sentence I am referring to. [. . .] When such a use of emphasis occurs neither the meaning nor the reference of the referring expression is altered with a change in emphasis. Accordingly, the sentence "Bill ate spoiled meat on Tuesday" = the sentence "Bill ate spoiled meat *on Tuesday*." Similarly, the sentence "Bill ate spoiled meat on Tuesday" = the sentence "Bill ate *spoiled meat* on Tuesday." Hence

(3) The sentence "Bill ate *spoiled meat* on Tuesday" = the sentence "Bill ate spoiled meat *on Tuesday*."

Suppose now that

(4) Dr. Smith explained Bill's stomach ache by uttering the sentence "Bill ate *spoiled meat* on Tuesday."

Dr. Smith emphasizes "spoiled meat" to indicate that he believes that this aspect of the situation is relevant for explaining Bill's stomach ache. He believes that it was the spoiled meat Bill ate that caused his stomach ache. Now if the u-position in sentences of the form "S explained q by uttering u" is referentially transparent, then from (4), in virtue of (3), we may infer

(5) Dr. Smith explained Bill's stomach ache by uttering the sentence "Bill ate spoiled meat *on Tuesday*."

But (5) is false since it entails that Dr. Smith believes that the day on which Bill ate the spoiled meat was relevant for his getting a stomach ache. And I am supposing that Dr. Smith had no such belief. His claim was that Bill got a stomach ache because of what he ate and not because of when he ate it. This distinction is expressed by the difference in emphasis between "The explanation of Bill's stomach ache given by Dr. Smith is that Bill ate *spoiled meat* on Tuesday," which is true if (4) obtains, and "The explanation of Bill's stomach ache given by Dr. Smith is that Bill ate spoiled meat *on Tuesday*," which is true if (5) obtains.

The emphases in (2) and

(6) The sentence "Bill ate *spoiled meat* on Tuesday"

are non-semantical. A shift of emphasis in these expressions will not change their meanings or references. But when these expressions are embedded in explanation-sentences such as (4) the emphases can assume a semantical role. The emphasized words can become "captured" by the term "explained," indicating that a particular aspect of Bill's eating spoiled meat on Tuesday is claimed to be explanatorily relevant for his getting a stomach ache. A shift in emphasis in (6) as this appears in (4) can transform (4), which is true, into (5), which is false.

This is not to deny that the emphases in both (4) and (5) could be used to play non-semantical roles. (One might assert (5) in response to someone who thinks that Dr. Smith explained Bill's stomach ache by uttering the sentence "Bill ate spoiled meat on Monday.") But if so (4) and (5) would have different readings from the ones being given here. We are using emphasis in (4) in the same way that Dr. Smith did, viz. to indicate a particular aspect of the situation Dr. Smith takes to be explanatorily relevant, not to correct a mistake about some of the words he uttered. When emphasis is understood as playing such a semantical role—which is possible in (4) and (5)—we get readings for these sentences under which (4) is true and (5) is false.

Assuming, then, that (2) and (6) denote the same sentence, and that the substitution of (2) for (6) in (4) turns a true sentence into a false one, we must conclude that the *u*-position in sentences of the form "*S* explains *q* by uttering *u*" is referentially opaque if that position is filled by sentence-expressions. There are readings for sentences of this form such that the substitution of co-referring *u*-terms will lead from truths to falsehoods. But the denotation condition of the proposition (as well as the sentence) view assumes that the *u*-position is transparent.

This problem can be avoided by introducing the notion of an *e*-sentence ("*e*" for emphasis), which is a sentence together with its emphasis, if any. To give an *e*-sentence that *S* uttered in explaining *q* one must supply *S*'s words, in the order he gave them, and with any emphasis he used. And we will now suppose that what is uttered by speakers in acts of explaining are *e*-sentences. The problem above is thus avoided since

(7) The *e*-sentence "Bill ate *spoiled meat* on Tuesday"

is not identical with

(8) The *e*-sentence "Bill ate spoiled meat *on Tuesday*."

Therefore from the fact that Dr. Smith explained Bill's stomach ache by uttering the *e*-sentence denoted by (7) it does not follow that he explained Bill's stomach ache by uttering the *e*-sentence denoted by (8).

Nevertheless there is a residual emphasis problem for the proposition view, which must now assume that *e*-sentences express propositions. Although the *e*-sentences denoted by (7) and (8) are not identical, the propositions these *e*-sentences express are. (7) and (8), respectively, denote the *e*-sentences

(9) Bill ate *spoiled meat* on Tuesday,

(10) Bill ate spoiled meat *on Tuesday*.

A shift in emphasis in (9) and (10) changes the *e*-sentence but not the proposition expressed. If I utter the *e*-sentence (9)—using emphasis where I do to indicate my surprise—and you think I am expressing the proposition that Bill ate spoiled meat on Monday, I may correct your mistake by uttering (10). By doing so I am not expressing a different proposition but the original one in a manner that will correct your misconception (that is, by using a different *e*-sentence).

Suppose then that Dr. Smith explained Bill's stomach ache by uttering the *e*-sentence (9), while Dr. Jones explained it by uttering the *e*-sentence (10). And in both cases assume that requirement (iii)—the uniqueness requirement—of the proposition theory's denotation condition is satisfied. It follows from this denotation condition that "the explanation of Bill's stomach ache given by Dr. Smith" denotes the same proposition as "the explanation of Bill's stomach ache given by Dr. Jones," viz. the proposition expressed by the *e*-sentences (9) and (10). Therefore

(11) The explanation of Bill's stomach ache given by Dr. Smith = the explanation of Bill's stomach ache given by Dr. Jones.

which we ought to reject since according to Dr. Jones's explanation, but not Dr. Smith's, the date on which Bill ate the spoiled meat is relevant for explaining Bill's stomach ache. If the date was irrelevant—if Bill would have gotten a stomach ache from eating spoiled meat on any day of the week—then, we may suppose, Dr. Smith's explanation is correct while Dr. Jones's is incorrect. We must conclude that these explanations are different and hence that (11) is false.

4. The Argument View

Before pursuing a remedy for these problems let me turn briefly to a different product view, suggested by the D-N model, according to which explanations are arguments.

Denotation condition (Argument view): "The explanation of q given by S" denotes an argument one of whose premises is (the proposition expressed by) u if and only if

(i) S explained q by uttering u;
(ii) $(v)(S$ explained q by uttering $v \supset v = u$ (or v and u express the same proposition)).

On one version of this view arguments are composed of propositions, and on the other of sentences. Since the proposition version will allow us to count explanations as identical even if they use different sentences, it will be considered. To ensure referential transparency in the u-position in "S explained q by uttering u" let us assume that the u's are e-sentences. Suppose, then, that Dr. Smith explained Bill's stomach ache by uttering the e-sentence

(1) Bill ate *spoiled meat* on Tuesday.

Assuming that condition (ii) above is satisfied, the product-expression

(2) "the explanation of Bill's stomach ache given by Dr. Smith"

denotes some argument one of whose premises is the proposition expressed by (1). What argument?

   Following the D-N model we may assume that it contains a law among its premises and that its conclusion, which is entailed by the premises, describes the event referred to in the object-expression q. We might suppose that the product-expression (2) denotes an argument composed of propositions expressed by sentences such as these:

(3) Bill ate *spoiled meat* on Tuesday. Anyone who eats spoiled meat gets a stomach ache. Hence, Bill got a stomach ache.

On the D-N view an argument such as (3) might be construed as an ordered set of propositions (or sentences), or perhaps better, as an ordered pair whose first member is the conjunction of premises and whose second is the conclusion.

   The argument view fares no better than the proposition view of Section 1. First, there is an illocutionary force problem. Poor Sam dies one hour after being operated on by Dr. Smith, who offers an excuse by uttering

(4) Sam had disease d at the time of his operation. Anyone with disease d at the time of an operation dies one hour after the operation. Hence, Sam died one hour after his operation.

Suppose further that Dr. Jones explains Sam's death by uttering (4). I will assume that, on the argument product-view, both "the excuse given by Dr. Smith" and "the explanation of Sam's death given by Dr. Jones" denote the argument given by (4), that is, the ordered pair whose first proposition is expressed by the conjunction of the first two sentences in (4) and whose second proposition is expressed by the last sentence. If so then the excuse given by Dr. Smith = the explanation of Sam's death given by Dr. Jones. But the excuse given by Dr. Smith was an excuse. Therefore,

(5) The explanation given by Dr. Jones was an excuse.

which is false if Dr. Jones has no intention of excusing anything. Indeed Dr. Jones may believe the operation and resulting death to be inexcusable. What is true is only that what Dr. Smith uttered in giving his excuse was what Dr. Jones uttered in explaining Sam's death. But (5) does not follow from that.

The emphasis problem can also be shown to be present, but I will omit the details.

5. Proposition Construed As Explanatory

A view will now be developed which avoids the illocutionary force problem. Later it will be shown how it can be modified to avoid the emphasis problem as well.

The first difficulty with the proposition (and the argument) view is that S may explain q by uttering what is uttered by someone performing a different type of illocutionary act, such as criticizing or excusing. If the product-expression denotes the proposition expressed by the sentence S utters in his act of explaining, then S's explanation can turn out to be a criticism or an excuse even when it is not. The proposition view fails to take proper account of the fact that to explain is to utter something with a certain illocutionary force. Perhaps then the product of an explanation should be taken to be what is expressed by that utterance—a proposition—construed with the illocutionary force of explaining.

Searle in a discussion of illocutionary acts criticizes Austin for failing to distinguish two senses of "statement": the act of stating and what is stated (which Searle calls the statement-object and I call the product of the act of stating).[9] He goes on to identify the latter as a "proposition construed as stated."[10] His discussion here is very brief and we are not told what sort of thing this is. But if we knew we could better understand products of explanation. That is, following Searle's discussion of statements, having distinguished the act of explaining from the product of the act, we now identify the latter as a proposition construed as explaining. But what sort of entity is that?

The proposal I shall now consider is that it is an ordered pair consisting of a proposition and a type of illocutionary act, in the present case an explaining type of act. So, for example,

(1) (The proposition that Bill ate spoiled meat; explaining Bill's stomach ache)

is a product of explanation, since it is an ordered pair whose first member is a proposition and whose second is the act-type explaining Bill's stomach ache. The product expression

(2) "The explanation of Bill's stomach ache given by Dr. Smith"

could be said to denote the ordered pair (1) provided that Dr. Smith explained Bill's stomach ache by uttering something, and whatever the doctor explained Bill's stomach ache by uttering expresses the proposition that Bill ate spoiled meat. More generally, we have the following

Denotation condition (preliminary ordered pair view): "The explanation of q given by S" denotes $(x;y)$ if and only if

    (i) x is a proposition;
    (ii) $y =$ (the act-type) explaining q;
    (iii) $(\exists u)(S$ explained q by uttering $u)$;
    (iv) $(u)(S$ explained q by uttering $u \supset u$ expresses $x)$.[11]

This view avoids the illocutionary force problem. The product-expression (2) denotes the ordered pair

(1) (The proposition that Bill ate spoiled meat; explaining Bill's stomach ache),

while

(3) "The criticism of Bill given by Jane"

denotes

(4) The proposition that Bill ate spoiled meat; criticizing Bill).

But (1) $\neq$ (4). Therefore, neither "The explanation of Bill's stomach ache given by Dr. Smith = the criticism of Bill given by Jane" nor "The explanation of Bill's stomach ache given by Dr. Smith was a criticism" can be derived. This view avoids the illocutionary force problem by taking into account not simply the proposition expressed in the act but also the type of illocutionary act itself. The types of illocutionary acts associated with the products denoted by (2) and (3) are different even though the propositions expressed are the same.

On the other hand, the emphasis problem remains, as can be seen if we suppose that Dr. Smith explained Bill's stomach ache by uttering "Bill ate *spoiled meat* on Tuesday," while Dr. Jones explained it by uttering "Bill ate spoiled meat *on Tuesday*." On the present view, the explanation given by each doctor is the ordered pair (The proposition that Bill ate spoiled meat on Tuesday; explaining Bill's stomach ache), since the *e*-sentences uttered by the doctors express the same proposition. Hence, the explanation of Bill's stomach ache given by Dr. Smith = the explanation of Bill's stomach ache given by Dr. Jones, which is false.

Can the ordered pair view be modified to avoid this problem? In the following section a proposal will be suggested for doing so.

6. A New Ordered Pair View

In Part I the conditions for S performing an act of explaining q by uttering u require that S intend that u express a correct answer to question Q. A view of explanation-products as answers to questions will now be explored.

To do so let us recall some concepts [. . .]. A sentence such as

(1) The reason that Bill got a stomach ache is that he ate spoiled meat

is a content-giving sentence for the noun "reason." It expresses a content-giving proposition (for the concept expressed by "reason"). Assuming that (1) expresses the same proposition as

(2) Bill got a stomach ache because he ate spoiled meat,

the proposition expressed by (2) is also a content-giving proposition. Furthermore, the propositions expressed by (1) and (2) are complete content-giving propositions with respect to the question

(3) Why did Bill get a stomach ache?

The proposition expressed by "Bill ate spoiled meat" is not a complete content-giving proposition with respect to (3), since a complete content-giving proposition with respect to Q entails all of Q's presuppositions.

Finally, question Q was called a content-question if and only if there is a complete content-giving proposition with respect to Q. Since (1) and (2) express complete content-giving propositions with respect to (3), the latter is a content-question. [. . .]

Returning now to the ordered pair view, my proposal is that the constituent proposition in the ordered pair is a complete content-giving proposition with respect to a certain question. This question will be given by the object-expression q. As in the case of "S explains q by uttering u" it will be supposed that any product-expression of the form "the explanation of q given by S" is, or is transformable into, one in which q is an indirect interrogative whose direct form expresses a content-question Q. Thus the product-expression "the explanation of Bill's stomach ache given by Dr. Smith" could be reformulated as "the explanation given by Dr. Smith of why Bill got a stomach ache." And the latter will be taken to denote an ordered pair one of whose constituents is a complete content-giving proposition with respect to question (3). But which proposition will this be and how is it related to Dr. Smith's act of explaining?

Recall from Part I the u-restriction imposed on explaining acts: what S utters is, or in the context is transformable into, a sentence expressing a complete content-giving proposition with respect to Q. If p is such a proposition, let us say that p is *associated with* S's explaining act. Equivalently, p is associated with S's act of explaining q by uttering u if

    (a) p is a complete content-giving proposition with respect to Q;
    (b) The act was one in which p was claimed to be true;
    (c) p entails the proposition expressed by u.[12]

According to these conditions, if Dr. Smith explained why Bill got a stomach ache, by uttering

(4) The reason that Bill got a stomach ache is that he ate *spoiled meat* on Tuesday.

then the proposition (4) is associated with the act of explaining performed by Dr. Smith. But equally if Dr. Smith explained why Bill got a stomach ache, by uttering

(5) Bill ate *spoiled meat* on Tuesday,

then proposition (4) is also associated with Dr. Smith's act of explaining, even though proposition (4) is not expressed by the *e*-sentence which Dr. Smith uttered. If by uttering (5) Dr. Smith explained why Bill got a stomach ache, his act of explaining was one in which the truth of proposition (4) was still being claimed. Since proposition (4) is a complete content-giving one with respect to question (3), and it entails the proposition expressed by sentence (5) which Dr. Smith uttered, proposition (4) is associated with his act of explaining.

Now the constituent proposition in the explanation of *q* given by *S* will be one *associated with S*'s act(s) of explaining *q*. Accordingly, we may formulate the following

Denotation condition (new ordered pair view): "The explanation of *q* given by *S*" denotes $(x;y)$ if and only if

 (i)   *Q* is a content-question;
 (ii) *x* is a complete content-giving proposition with respect to *Q*;
 (iii) *y* = explaining *q*;
 (iv) $(\exists a)(\exists u)(a$ is an act in which *S* explained *q* by uttering *u*);
 (v) $(a)(u)[a$ is an act in which *S* explained *q* by uttering $u \supset (r)(r$ is associated with $a \equiv r = x)]$. (That is, *x* is the one and only proposition associated with every act in which *S* explained *q* by uttering something.)[13]

Since (4) is a complete content-giving proposition with respect to the content-question (3), "the explanation given by Dr. Smith of why Bill got a stomach ache" denotes (proposition (4); explaining why Bill got a stomach ache), provided that there was an act in which Dr. Smith, by uttering something, explained why Bill got a stomach ache, and for any such act proposition (4) was the one and only associated proposition.

Returning now to the definition of association, why are such propositions and only these taken to be constituent propositions of explanations? That is, why are conditions (a) through (c) for association imposed? The first, by requiring that *p* be a complete content-giving proposition with respect to the content-question *Q*, guarantees that *p* is an answer to *Q*; and this relates the product of explanation to the act in which the explainer intends to provide an answer to *Q*. Moreover, by requiring that answer to *Q* to be a complete content-giving proposition the emphasis problem is avoided (as demonstrated in the next section). The reason for condition (b) can be shown by an example in which Dr. Smith by uttering the *e*-sentence

(5) Bill ate *spoiled meat* on Tuesday

explains why Bill got a stomach ache. The proposition

(6) The reason Bill got a stomach ache is that Bill ate spoiled meat *on Tuesday*

is a complete content-giving one with respect to question (3), and it entails the proposition expressed by the *e*-sentence Dr. Smith uttered, viz. (5). But (6) could not be the constituent proposition in Dr. Smith's explanation. (Otherwise his explanation would be identical with Dr. Jones's, which it isn't.) Condition (b) precludes this on the grounds that (6) is not a proposition claimed to be true in Dr. Smith's act of explaining. By uttering (5) in explaining why Bill got a stomach ache Dr. Smith does not claim the truth of (6).

The reason for condition (c) can also be given by means of the above example in which Dr. Smith by uttering (5) explained why Bill got a stomach ache. The proposition

(7) The reason Bill got a stomach ache is that he ate spoiled meat

is a complete content-giving one with respect to question (3), and it is also a proposition claimed to be true in Dr. Smith's act of explaining. Both propositions (4) and (7) satisfy conditions (a) and (b), and without condition (c) there would be no unique proposition to associate with Dr. Smith's explaining act. Condition (c) precludes (7) as a proposition associated with this act, since (7) does not entail the proposition expressed by the *e*-sentence Dr. Smith uttered, viz. (5).

In choosing the constituent proposition to be one satisfying the conditions of *association,* the product of explanation is seen to be intimately related to the act of explaining. The constituent proposition is an answer to a question (a question which the explainer in his act intends to answer); it is one claimed to be true in the explaining act; and either it is one expressed by what the explainer utters in that act or it entails the proposition expressed by what the explainer utters.

The new ordered pair view can also supply conditions for being *an* explanation of q given by S. Here since uniqueness is not required, we need not suppose that the same proposition is associated with every act in which S explained q. Accordingly, we can write

$(x;y)$ is an explanation of q given by S if and only if

    (i) Q is a content-question;
    (ii) x is a complete content-giving proposition with respect to Q;
    (iii) $y = $ explaining q;
    (iv) $(\exists a)(\exists u)(a$ is an act in which S explained q by uttering u, and x is associated with a).

As will be shown in Section 8, the ordered pair view can be generalized to explanations that are not given by anyone (i.e., that are not products of any explaining acts). What first needs to be asked, however, is whether it is subject to the difficulties of the previous views.

7. Are the Illocutionary Force and Emphasis Problems Avoided?

Since the second constituent of the product of explanation is an explaining type of act we avoid the illocutionary force problem. Even if the constituent proposition of Dr. Smith's explanation is the same as that of Jane's criticism it will not follow that "Dr. Smith's explanation of why Bill got a stomach ache" and "Jane's criticism of Bill" denote the same entity.

Second, if the constituent proposition of an explanation is a complete content-giving one with respect to Q then the emphasis objection is avoided. Suppose that the explanation given by Dr. Smith of why Bill got a stomach ache is that Bill ate *spoiled meat* on Tuesday, and that the explanation given by Dr. Jones is that Bill ate spoiled meat *on Tuesday*. The emphasis problem arises if we take the constituent propositions of these explanations to be ones expressed by the *e*-sentences

    (1) Bill ate *spoiled meat* on Tuesday,

    (2) Bill ate spoiled meat *on Tuesday*.

The propositions expressed by (1) and (2) are identical, from which we would have to conclude that the explanations given by the two doctors are identical. However, the propositions expressed by (1) and (2) are not complete content-giving ones with respect to "Why did Bill get a stomach ache?" Therefore, on the ordered pair view of Section 6 these propositions are not the constituent propositions of the explanations in question. But we can suppose that the constituent proposition of the explanation given by Dr. Smith is

    (3) The reason Bill got a stomach ache is that he ate *spoiled meat* on Tuesday,

while that of the explanation given by Dr. Jones is

    (4) The reason Bill got a stomach ache is that he ate spoiled meat *on Tuesday*.

(3) and (4) are complete content-giving propositions with respect to "Why did Bill get a stomach ache?"; and these propositions are not identical. Shifting emphasis in (3) and (4), unlike (1) and (2), means changing the propositions expressed. If Dr. Smith by uttering (1) explained why Bill got a stomach ache, he put the emphasis where he did to indicate an explanatorily relevant aspect of an event. But the emphasis in (1) does not indicate this. By contrast, the emphasis in (3) does, since it is captured by "reason." In general, in a complete content-giving proposition any explanatory emphasis used by the explainer will be so captured.

    Why is this so? Consider a content-giving sentence of form (i) [. . .] (any content-giving sentence is equivalent to one of this form):

(i) $\begin{Bmatrix} \text{The} \\ \text{A} \end{Bmatrix}$ + content-noun N + phrase + to be + (preposition) + nominal.

Let us also consider a question of the form (or equivalent to one of the form)

    What + is + the + content-noun N + phrase?

which is constructed from (i). Now instead of answering a question of this form by uttering a sentence of form (i), one might answer simply by uttering the nominal in (i). (If the nominal is a that-*p* clause, one might simply utter *p*.) For example, suppose that the question is

    Q: What is the reason that Bill got a stomach ache?

or, its equivalent, "Why did Bill get a stomach ache?" In explaining *q*, instead of uttering (something equivalent to)

    (5) The reason that Bill got a stomach ache is that Bill ate *spoiled meat* on Tuesday,

which has form (i), someone might utter

(1) Bill ate *spoiled meat* on Tuesday,

which is just the *p*-sentence in the nominal in (i). But when a speaker by uttering (1) explains why Bill got a stomach ache, he is implicitly claiming that (5) is true. (The proposition expressed by (5) is associated with his act of explaining.)

Now in (5) the emphasized words are captured by the content-noun "reason." In general, [. . .] many (though not all) content-nouns in content-giving sentences of form (i)—as well as in content-giving sentences of other forms—are emphasis-selective: in sentences of form (i), emphasized words in the nominal become selected or captured by the content-noun, indicating that a particular aspect of the situation described by the nominal is operative (is the reason, the cause, what is important, etc.). A shift of emphasis, as from (3) to (4) above, results in a shift in the operative aspect; and this can change the truth-value of the resulting sentence. If the content-noun is emphasis-selective (as is "reason"), then a content-giving sentence of form (i), or any equivalent content-giving sentence, will be able to incorporate operative aspects emphasized by explainers who may utter abbreviated versions of such sentences.

Thus, when Dr. Smith utters the abbreviated (1) in explaining why Bill got a stomach ache, he is selecting the item that Bill ate as an explanatorily operative feature. This feature is explicitly selected by the content-noun "reason" in the sentence (3) which expresses a proposition associated with Dr. Smith's act of explaining. The proposition expressed by (3) is a complete content-giving one with respect to the question "Why did Bill get a stomach ache?" More generally, in an explaining act a speaker may employ explanatory emphasis in such an abbreviated sentence. This emphasis can be explicitly captured in a sentence that expresses a complete content-giving proposition that is associated with this explaining act.

8. A Generalization of the Ordered Pair View and Comparisons with Other Theories

Expressions of the form "the explanation of *q* given by *S*" have been said to denote something only if *S* has explained *q*. But we can also refer to explanations that are not products of any acts of explaining—explanations no one has given—using expressions of the forms "the explanation of *q*" and "the explanation of *q* that is *F*." The use of the term "product-expression" will be extended to cover expressions of these forms which may denote entities even if no explaining acts have been performed. We might say that they denote entities that if not actually products of explaining acts are so potentially. Can the ordered pair view provide denotation conditions for this more general class of product-expressions?

The only difference between "the explanation of *q*" and "the explanation of *q* given by *S*" is that the use of the latter but not the former entails that there was some act in which *S* explained *q*. This entailment is reflected in (iv) of the ordered pair denotation condition for "the explanation of *q* given by *S*." If (iv) is dropped and (v) is suitably generalized so as not to refer to a particular explainer, we obtain

Denotation condition: "The explanation of *q*" denotes $(x;y)$ if and only if

    (i)   *Q* is a content-question;

    (ii) *x* is a complete content-giving proposition with respect to *Q*;

(iii) y = explaining q;

(iv) $(a)(S)(u)$ [a is an act in which S explains q by uttering $u \supset (r)(r$ is associated with $a \equiv r = x)$].[14]

(iv), which expresses a uniqueness condition, does not require that any particular act of explaining q has occurred but only that if any does then x is the only proposition associated with it. To obtain a denotation condition for "the explanation of q that is F" (e.g., "the explanation of q that is the least plausible") we retain (i), (ii), and (iii) above, replace (iv) with

  $(a(S)(u)$ (a is an act in which S explains q by uttering u, and x is the one and only proposition associated with a. $\supset a$ is an act in which S explains q in the F-manner),

and add

  $(a)(S)(a$ is an act in which S explains q in the F-manner $\supset a$ is an act for which x is the one and only associated proposition).

The ordered pair theory can provide conditions for sentences of the form

  (1) An explanation of q is that c.

Let us call a sentence *restructured* if q is an indirect interrogative expressing a content-question Q and c is a sentence expressing a complete content-giving proposition with respect to Q. I shall assume that any unrestructured sentence of form (1) is paraphrasable into a restructured sentence of this form. For example,

  An explanation of why Bill got a stomach ache is that Bill got a stomach ache because he ate spoiled meat

is a restructured paraphrase of

  An explanation of Bill's stomach ache is that Bill ate spoiled meat.

On the ordered pair view, a restructured sentence of form (1) is true if and only if

  (2) $(\exists x)(x$ = (the proposition expressed by c; explaining q)).

  We can also say that

  (3) E is *an* explanation of q if and only if
    (i) Q is a content-question;
    (ii) E is an ordered pair whose first member is a complete content-giving proposition with respect to Q and whose second member is the act-type *explaining q*.

For $(x;y)$ to be *an* explanation of q we need not suppose that any particular explaining acts have occurred. Nor must we suppose, as was done earlier to ensure uniqueness, that in any act in which q is explained the constituent proposition x is associated with that act. On the present view, an explanation of q is an ordered pair whose second member is the act type "explaining q" and whose first member is an answer to Q that is a complete content-giving proposition with respect to Q.

It is at this point that the present account of explanation-products is most use-fully compared with several [others]. Aristotle and Hempel, for example, can be construed as providing conditions not for E's being *someone's* explanation of q, but for E's being *an* explanation of q, as follows:

Aristotle: E is an explanation of q if and only if

    (i) Q is a question of the form "Why does X have P?"

    (ii) E is a proposition of the form "X has P because r" in which r pur-ports to give one or more of Aristotle's four causes of X's having P.

Hempel: E is an explanation of q if and only if

    (i) Q is a question of the form "Why is it the case that p?"

    (ii) E is a D-N (or inductive) argument, whose conclusion is p, that contains lawlike sentences (and that satisfies other conditions Hem-pel imposes).

It can readily be seen that (3) above provides much broader conditions than either of these. (3) allows questions associated with explanations to be content-ques-tions of any sort, and not simply content-questions of the form "Why does X have P?" (Aristotle), or "Why is it the case that p?" (Hempel), or "Why is that X, which a member of A, a member of B?" (Salmon). Explanations answer a variety of ques-tions of forms other than these.

Second, allowing the constituent proposition of an explanation of q to be any complete content-giving proposition with respect to Q permits us to identify as ex-planations a variety not recognized by the above accounts. There is no need to restrict constituent propositions of explanations to ones of the form "X has P because r" in which r gives a cause, or "p, therefore r," where Hempel's conditions are satisfied. For example, the following count as explanations, according to (3), even if their constituent propositions are not expressible by sentences of these forms:

> (The event now occurring in the bubble chamber is that alpha particles are passing through; explaining what event is now occurring in the bubble chamber);

> (The significance of that document is that it is the first to proclaim the rights of animals; explaining what the significance of that document is);

> (The purpose of the flag is to warn drivers of danger; explaining what the purpose of that flag is).

Finally, and very importantly, without associating illocutionary act-types with explanations, as is done in the second condition of (3)—or at least without some such proviso—we have been unable to distinguish explanations from products of other (possible) illocutionary acts. For example, what makes the proposition "Bill got sick because he ate spoiled meat" an explanation rather than an excuse, or a complaint, or a simple identification of who got sick because he ate spoiled meat? Similar ques-tions are possible with respect to arguments of the form "p, therefore q." This is the illocutionary force problem discussed in Section 2. What the ordered pair theory is proposing is that to distinguish an explanation from any one of a number of other illocutionary products we invoke illocutionary acts (i.e., types). The propositions expressed in such acts will not suffice to distinguish the products, since the propositions may be identical even though the products are not. The conditions in (3) make essential reference to a type of explaining act, viz. explaining q, where q is whatever it is that is being explained. [. . .] By contrast, theories of explanation [. . .] such as Aristotle's and Hempel's—provide conditions for being an explanation that do not invoke the concept of an explaining act. Such theories do not adequately distinguish explanations from other illocutionary products.

. . . .

10. A No-Product View

A basic assumption of our discussion has been that explanation products are entities. And the most successful theory so far is the ordered pair view of Section 6. But an ontological purist may wonder whether he needs to postulate ordered pairs, or indeed anything else, as referents of explanation product-expressions. He may ask whether a sentence containing "the explanation of q given by S" can be understood without presupposing such entities but only particular acts of explaining. In this section a "reductionist" view will be considered according to which

> Any sentence containing an explanation product-expression is paraphrasable into a sentence that contains no such expression, but does contain one or more expressions of the form "a is an act in which S explained q by uttering u."

When this thesis is combined with the view that such paraphrases are more fundamental than sentences with product-expressions we get the no-product view.

To see how the paraphrases are supposed to work let me begin with sentences of the form

(1) The explanation of q given by S is that c.

Recalling a term introduced in Section 8, such a sentence is *restructured* if q is an indirect interrogative expressing a content-question Q and c is a sentence expressing a complete content-giving proposition with respect to Q. It is assumed that any unrestructured sentence of form (1) is paraphrasable into a restructured sentence of this fo.m. A restructured sentence of form (1) can then be understood as

(2) $(\exists a)(\exists u)(a$ is an act in which S explained q by uttering $u)$ and $(a)(u)(a$ is an act in which S explained q by uttering $u \supset (r)(r$ is associated with $a \equiv r =$ the proposition expressed by $c))$,

that is, there is an act in which S explained q by uttering something, and for any such act c expresses the one and only associated proposition. To understand (2) we need to suppose that S performed one or more explaining acts, but no products of those acts need to be postulated.

A restructured sentence of the form "*An* explanation of q given by S is that c" can be paraphrased simply as "$(\exists a)(\exists u)(a$ is an act in which S explained q by uttering u, and c expresses a proposition associated with $a)$," in which no product of explanation is invoked.

On the ordered pair view a sentence of the form

(3) The explanation of q given by $S_1 =$ the explanation of q given by S_2

is to be understood as a genuine identity in which the expression on the left is being said to denote the same product as the one on the right. On the no-product view (3) is not a genuine identity but is to be understood as

$(\exists a)(\exists b)(\exists u_1)(\exists u_2)(a$ is an act in which S_1 explained q by uttering u_1, and b is an act in which S_2 explained q by uttering u_2), and $(\exists p)(a)(b)(u_1)(u_2)(a$ is an act in which S_1 explained q by uttering $u_1 \supset p$ is the one and only proposition associated with a; and b is an act in which S_2 explained q by uttering $u_2 \supset p$ is the one and only proposition associated with b).

A sentence such as "The explanation of q given by S is correct" can be interpreted as

$(\exists a)(\exists u)(a$ is an act in which S explained q by uttering u) and $(\exists p)(a)(u)[a$ is an act in which S explained q by uttering $u \supset (p$ is the one and only proposition associated with a, and p is a correct answer to Q)].

That is, there is an act in which S explained q by uttering something, and there is a proposition uniquely associated with all such acts, and that proposition is a correct answer to Q.

Finally, how can we understand a restructured sentence of the form

(4) An explanation of q is that c

in which no uniqueness is assumed and no reference to a particular explainer is made? One way is to construe (4) as asserting the possibility of an act in which someone explains q by uttering something that expresses the proposition expressed by c, that is,

(5) Possible $(\exists a)(\exists S)(\exists u)(a$ is an act in which S explains q by uttering u, and u and c express the same proposition).

(Compare this with (2) of Section 8, which is proposed by the ordered pair view.) A stronger way of interpreting (4) above is as saying that a *correct* explanation of q is that p, which can be construed as a conjunction of (5) with "c is true" (provided that (4) is restructured) [. . .].

The present no-product view and the ordered pair view of Section 6 differ over whether explanations are entities. According to the no-product view, a restructured sentence of the form (1) is to be understood as a conjunction of an existentially general and a universally general sentence describing acts of explaining, viz. (2). According to the ordered pair view, (1) is a singular sentence in which the product-expression denotes an ordered pair. Nevertheless, there is an interesting similarity between these views. On the ordered pair view a restructured sentence of form (1) will be true if and only if the product-expression in (1) denotes an ordered pair consisting of the proposition expressed by c and the act-type explaining q, and (2) above obtains. (See the denotation condition of Section 6.) On the no-product view a restructured sentence of form (1) is paraphrased into (2). So what is a paraphrase of (1) on the no-product view is one of the truth-conditions for (1) on the ordered pair view.

Is one of these views superior? One consideration is how well each can justify inferences that are drawn from sentences containing explanation product-expressions. For example, from a sentence of form (1) we may quite naturally infer "S explained *q* by uttering something." Such an inference is sanctioned by the no-product view which paraphrases (1) as (2), which entails

(6) $(\exists a)(\exists u)(a$ is an act in which S explained q by uttering u).

But the ordered pair view is equally successful, since (6) is derivable from (1) via the denotation condition for "the explanation of *q* given by *S*." More generally, it can be shown that for every inference sanctioned by one view there is an identical or closely corresponding one sanctioned by the other.

Is one of these views preferable on ontological grounds? Both the ordered pair and no-product views invoke acts of explaining and propositions as entities. But the former in addition requires types of acts and ordered pairs consisting of these and propositions. Ontologists who crave simplicity may give an edge to the no-product view, provided that paraphrases such as those above are adequate. On the other hand, the no-product paraphrase of (4), viz. (5), uses a modal operator, which some may wish to avoid. In any case, a more liberal ontological position is possible. Two views may be equally satisfactory, even if one invokes more entities than the other, provided there are no special problems with these entities. Liberal ontologists who find types of acts no more problematic than propositions, and who have no reasons for rejecting ordered pairs or modal operators, may well conclude that there is little to choose between the ordered pair and no-product views. Finally, both views require the concept of an explaining act. Neither characterizes explanations independently of such a concept.

11. Implications for Standard Theories of Explanation

If one holds either the ordered pair view of Section 6 or the no-product view outlined in the preceding section, then at least parts of [earlier] theories of explanation [. . .] must be rejected. These theories—as I have been interpreting them—are product theories: they assert that explanations are entities of certain sorts—propositions or arguments. The no-product view rejects this idea. And the ordered pair view, while it accepts the idea that explanations are entities, refuses to identify them as the entities that Aristotle, Hempel, and others suggest. However, these philosophers make not only ontological claims about explanations but evaluative ones as well. Do the evaluative parts of their theories depend on the ontological ones? I suggest that they do not, for it is possible to retain their evaluative claims within a more successful ontological view.

To see this, it will suffice to consider Hempel's D-N theory. It is possible to reformulate this theory so that it will make claims not about the ontological character of explanations but only about their goodness. On this construction, the D-N theorist could accept the ordered pair view of Section 6. He could say that an explanation is an ordered pair $(p;$ explaining $q)$ whose first member is a complete content-giving proposition with respect to Q and whose second member is the act-type explaining q, where Q is a content-question. His theory of evaluation will now be restricted to explanations of certain types only, as follows.

D-N condition of evaluation (using ordered pair view): If Q is a content-question expressible by a sentence of the form "Why is it the case that *x*?" and $(p;$ explaining *q*) is an explanation of *q* in which proposition *p* is expressible by a sentence of the form "The reason that *x* is that *y*," then (*p*; explaining *q*) is a *good explanation of q* if and only if

(i)   *y* is a true sentence;
(ii) *y* is a conjunction at least one of whose conjuncts is a law;
(iii) *y* entails *x*.

For example, by this condition the ordered pair

> (The reason that this metal expanded is that this metal was heated and all metals expand when heated; explaining why this metal expanded)

is a good explanation of why this metal expanded.

Once the D-N theory is reformulated as a theory without particular ontological commitments but only as a theory for evaluating explanations, it is also possible to use the D-N theory in connection with the no-product view:

D-N condition of evaluation (using no-product view): If Q is a content-question expressible by a sentence of the form "Why is it the case that *x*?" then the explanation of *q* given by *S* is good if and only if $(\exists a)(\exists u)(a$ is an act in which *S* explained *q* by uttering *u*), and $(\exists p)(a)(u)(\exists x)(\exists y)(a$ is an act in which *S* explained *q* by uttering $u \supset p$ is the one and only proposition associated with *a*, and *x* and *y* are sentences, and the sentence "the reason that *x* is that *y*" expresses *p*, and *y* is a conjunction at least one of whose conjuncts is a law, and *y* entails *x*).

The D-N theory, as formulated by Hempel, has quite definite ontological commitments. These are incompatible with the ordered pair and no-product views. However, it is possible to reformulate the D-N theory so that it provides conditions only for the evaluation of explanations. So reformulated, it is neutral between product and no-product views. [. . .] This is not to say, of course, that the resulting evaluative theories are acceptable. [. . .]

12. Conclusions

On the usual views, explanations are entities of certain sorts—sentences, propositions, or arguments—that are to be characterized independently of the concept of an explaining act. These views are subject to the illocutionary force problem: they are unable to distinguish explanations from other illocutionary products. They are also subject to the emphasis problem: they are unable to capture the idea that shifts of emphasis in what is asserted in an explanation can change the meaning of what is said, and hence the identity of the explanation. These problems are avoided by the ordered pair and no-product views, both of which appeal to the concept of an explaining act. One holding either of these views must reverse the usual order of procedure. Such a person must begin with the concept of an illocutionary act of explaining and characterize explanations, by reference to this, rather than conversely. Finally, we saw how the evaluative aspect of a typical theory of explanation—the D-N model—can be retained while abandoning its ontological commitments in favor of those of the ordered pair and no-product views.

Notes

1. Zeno Vendler, *Linguistics in Philosophy* (Ithaca, 1967), p. 102.

2. Sylvain Bromberger, "An Approach to Explanation," in R. J. Butler, ed., *Analytical Philosophy* 2 (Oxford, 1965), pp. 72–105.

3. Donald Davidson, "The Logical Form of Action Sentences," in N. Rescher, ed., *The Logic of Decision and Action* (Pittsburgh, 1967).

4. J. L. Austin, *How To Do Things with Words* (Oxford, 1962). Austin includes "explain" on his list of "expositives," pp. 160–61.

5. The product-expression "the explanation of q given by S" may be used to refer uniquely even if S gave different explanations of q; for example, when we use this expression we might be referring to the explanation of q given by S during time t, in which case the expression "S explained q by uttering u" in (ii) and (iii) becomes "S explained q by uttering u during t." Again, the product-expression may be used to refer uniquely even if S explained q on many occasions provided he gave the same explanation on each.

6. It could also be avoided by identifying explanations with *sets* of sentences that are equivalent in meaning. This is subject to the same difficulties as the proposition view to be considered next.

7. The denotation condition would be this: "the criticism given by S" denotes x if and only if (i) x is a proposition, (ii) $(\exists u)(S$ criticized by uttering $u)$, (iii) $(u)(S$ criticized by uttering $u \supset u$ expresses $x)$.

8. For a seminal discussion of emphasis, see Fred Dretske, "Contrastive Statements," *Philosophical Review* 81 (1972), pp. 411–37.

9. John Searle, "Austin on Locutionary and Illocutionary Acts," *Philosophical Review* 77 (1968), pp. 405–24.

10. Ibid., p. 423.

11. Analogous conditions are possible for other illocutionary product-expressions such as "the criticism of q given by S."

12. If u is not a sentence, then, by the u-restriction, it is transformable into one that is; and p will entail the proposition expressed by the latter. In what follows only u's that are sentences will be considered.

13. Given the definition of "association," (i) and (ii) are redundant since they are entailed by (iv) and (v).

14. "The explanation of q" in which no further clause is added is often used to mean "the correct explanation of q." In such a case "explains" in condition (iv) can be understood as "correctly explains."